Cryogenic Microelectronic Systems for Ultra-Low Energy and Enhanced Performance

Nurzhan Zhuldassov • Eby G. Friedman

Cryogenic Microelectronic Systems for Ultra-Low Energy and Enhanced Performance

Springer

Nurzhan Zhuldassov
Electrical and Computer Engineering
University of Rochester
Rochester, NY, USA

Eby G. Friedman
Electrical and Computer Engineering
University of Rochester
Rochester, NY, USA

ISBN 978-3-031-90753-1 ISBN 978-3-031-90754-8 (eBook)
https://doi.org/10.1007/978-3-031-90754-8

This Springer imprint is published by the registered company Springer Nature Switzerland AG
The registered company address is: Gewerbestrasse 11, 6330 Cham, Switzerland

For my parents, Zamzagul and Gazez, whose unwavering love, selfless sacrifices, and steadfast support have been the foundation of all I do.

Nurzhan

To my wife, sons, daughters-in-law, and granddaughters.

Eby

Preface

The primary purpose of this book is to introduce and discuss the design and optimization of computing systems operating at cryogenic temperatures. The topic is inspired by the realization that conventional transistor scaling has approached fundamental physical limits. The existing challenges of heat density and quantum tunneling effects in sub-three nanometer technology nodes have necessitated the exploration of alternative methods for sustaining and extending computational capabilities.

Cryogenic computing represents a promising path forward. Operating electronic systems at cryogenic temperatures offers several fundamental advantages in semiconductors, including enhanced carrier mobility, reduced thermal noise, faster switching speeds, and negligible leakage currents. The impact of these benefits extend to a wide spectrum of applications, most notably quantum computing and high performance cloud computing systems. Integration of cryogenic CMOS with quantum processors facilitates larger qubit arrays by minimizing wiring overheads and thermal gradients. Furthermore, negligible leakage currents enable efficient dynamic logic that is often infeasible at room temperature and low frequencies. Despite the power cost of cooling, these benefits can produce notable gains in system-wide energy efficiency, particularly for stationary computing platforms such as cloud infrastructure and quantum computers. The topics presented in this book address issues that have not been examined in other publications; namely, how to determine which technology and which circuit function should operate in which temperature zone. A novel graph theoretic methodology is introduced to optimize system components across multiple temperature domains, balancing performance requirements with energy consumption and the cooling infrastructure.

This monograph is based on doctoral research undertaken by Nurzhan Zhuldassov from 2018 to 2024 at the University of Rochester under the supervision of Professor Eby G. Friedman. The book offers a cohesive understanding of cryogenic computing by merging fundamental circuit principles at low temperatures with system level optimization techniques. The chapters cover a range of topics, from transistor behavior at low temperatures and multi-temperature thermal models to advanced design approaches for dynamic logic circuits operating at cryogenic temperatures. Methodologies are presented through case studies of cloud and quantum computing, demonstrating the potential for significant power savings despite the energy cost of cooling.

Several chapters contain entirely original work, including the introduction of a novel optimization procedure that incorporates both circuit performance and system-level thermal constraints. The importance of this issue becomes apparent in heterogeneous cryogenic systems, where multiple technologies may require distinct operating temperatures to achieve optimal results. By establishing a framework for systematically choosing different operating temperatures, the material presented here addresses a gap in competing works that often limit attention to a single temperature zone or utilize an *ad hoc* approach to choose a specific temperature.

This book aims to serve as a valuable resource for industrial system designers, academic researchers, and advanced graduate students seeking to understand and develop cryogenic computing systems. A discussion of system design methodologies, the exploration of real world case studies, and an emphasis on future directions underscore the significance of cryogenic computing in an era where traditional scaling trends can no longer provide the desired gains in performance and energy efficiency. It is anticipated that these insights will contribute to further innovations in high performance low temperature computing, paving the way for more widespread use of cryogenic systems in next generation electronics.

Rochester, NY, USA

Nurzhan Zhuldassov
Eby G. Friedman

Acknowledgments

The authors would like to thank the Editorial Director of Springer, Charles Glaser, for his help and assistance. His support over the years is deeply appreciated.

The original research work presented in this monograph was made possible in part by the National Science Foundation under Grant No. 2124453 (DISCOVER Expeditions), Grant No. 2308863 (Cryogenic Hybrid Systems Integration Across Multiple Temperature Zones), and Grant No. CCF-1716091, Intelligence Advanced Research Projects Activity under Grant No. W911NF-17-9-001, the Department of Energy under FOA number DE-FOA-0002950, and by grants from Cisco Systems, Qualcomm, Synopsys, and Google.

Competing Interests The author has no competing interests to declare that are relevant to the content of this manuscript.

Contents

About the Authors

Nurzhan Zhuldassov received his B.S. degree in Electrical and Computer Engineering from Nazarbayev University, Astana, Kazakhstan in 2018, and his M.S. in 2019 and Ph.D. in 2024, both in Electrical and Computer Engineering at the University of Rochester, Rochester, New York.

He was a hardware engineering intern at Google, Sunnyvale, California, and he was a hardware engineering intern at Apple, Cupertino, California. He is currently employed by Apple. His research interests encompass on-chip and package level power delivery networks, power integrity, cryogenic operation of MOSFETs, and optimization of cryogenic electronic circuits. He has authored 16 articles.

Eby G. Friedman (Life Fellow, IEEE) received the B.S. degree in electrical engineering from Lafayette College, and the M.S. and Ph.D. degrees in electrical engineering from the University of California at Irvine. He was with Hughes Aircraft Company, from 1979 to 1991, rising to Manager of the Signal Processing Design and Test Department, where he was responsible for the design and test of high performance digital and analog ICs. He has been with the Department of Electrical and Computer Engineering, University of Rochester, since 1991, where he is currently a Distinguished Professor and the Director of the High Performance VLSI/IC Design and Analysis Laboratory. He is also a Visiting Professor with the Technion-Israel Institute of Technology, Haifa, Israel. He has authored more than 600 articles and book chapters, holds 29 patents, and has authored or edited 21 books in the fields of high speed and low power CMOS design techniques, 3-D design methodologies, high speed interconnect, superconductive circuits, and the theory and application of synchronous clock and power distribution networks. His current research and teaching interests include high performance synchronous digital and mixed-signal circuit design and analysis with application to high speed portable processors, low power wireless communications, and large scale data centers. Dr. Friedman is a Senior Fulbright Fellow, a Fellow of the NAI, a National Sun Yat-sen University Honorary Chair Professor, and an Inaugural Member of the UC Irvine Engineering Hall of Fame. He was a recipient of the IEEE Circuits and Systems Mac Van Valkenburg Award, the IEEE Circuits and Systems Charles A. Desoer Technical Achievement Award, the University of Rochester Graduate Teaching Award and Lifetime Achievement Award, and the College of Engineering Teaching Excellence Award. He was the Editor-in-Chief and Chair of the Steering Committee of the *IEEE Transactions on Very Large Scale Integration (VLSI) Systems*, the Editor-in-Chief of the Microelectronics Journal, a Regional Editor of the *Journal of Circuits, Systems and Computers*, an editorial board member for numerous journals, and a program and technical chair for several IEEE conferences.

Chapter 1
Cryogenic Principles and Applications

Cryogenics, the field dedicated to the production and maintenance of extremely low temperatures, is crucial across a broad spectrum of applications. These cryogenic applications include electronics, medical imaging, food preservation, military systems, quantum computing, artificial intelligence, and high performance computing. These diverse areas of application are depicted in Fig. 1.1.

Cryogenics significantly influences several sectors including engineering, science, medicine, and economics. A study from the UK highlighted that in 2015 17% of the UK economy was linked to cryogenic technology [1]. In scientific research, the reduction of molecular vibrations at cryogenic temperatures can be advantageous, including microscopy where less motion aids in visualizing molecular structure through neutron scattering [2, 3]. This property is also beneficial in astronomy, where cryogenically cooled detectors reduce signal noise [4].

In industry, cryogenics supports environmental sustainability, particularly in the natural gas sector. Cryogenics is used to transform natural gas into liquefied natural gas (LNG), greatly reducing volume for improved transportation efficiency [5]. In the energy storage domain, cryogenics enables the storage of surplus energy by liquefying gases, later used to generate electricity [6].

In medical settings, cryogenics facilitates the preservation of biological and organic materials, enabling long term access to tissue samples. Hospitals also rely on cryogenic storage of liquid oxygen tanks for patient care. A notable use of medical imaging is magnetic resonance imaging, where liquid helium at 4 K sustains superconducting magnets. Beyond healthcare, superconductive magnets are essential in particle colliders like the Large Hadron Collider at CERN [7] and the international thermonuclear experimental reactor [8], where magnets contain plasma fusion fuel using niobium-tin superconductors to generate extremely high magnetic fields.

Cryogenics also plays a critical role in sensors in many fields, ranging from missile tracking and electrical machines [9] to advanced IR and RF sensors [10]. This technology is poised to enhance photonic, electro-optic, and optoelectronic devices

N. Zhuldassov, E. G. Friedman, *Cryogenic Microelectronic Systems for Ultra-Low Energy and Enhanced Performance*, https://doi.org/10.1007/978-3-031-90754-8_1

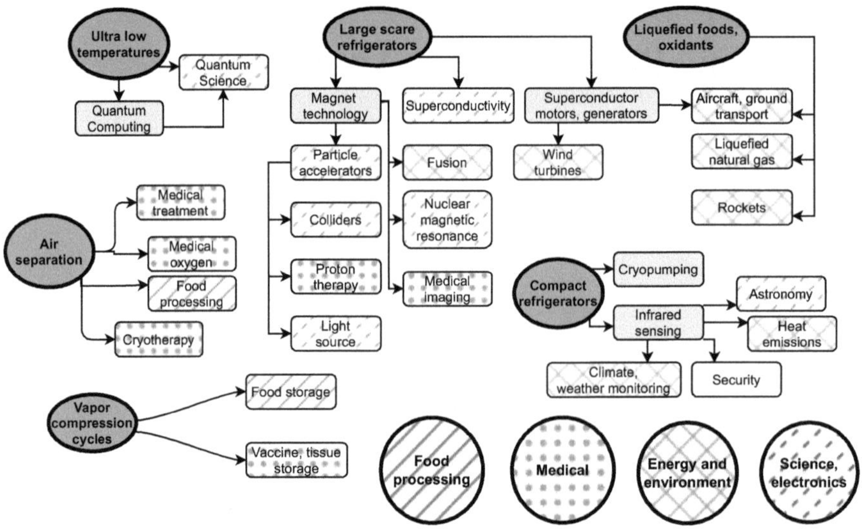

Fig. 1.1 Different fields utilizing cryogenics

[10], illustrating the expansive impact of cryogenics across multiple domains. Finally and highly relevant to this book, cryogenics plays an important role in computing, allowing for fast and energy efficient operation of computing devices. In particular, the stationary nature of cloud computing systems enables the use of cryogenic cooling.

The history of cryotechnology and cryogenic cooling is described in Section 1.1. The benefits and applications of cryogenics are discussed in Section 1.2. A brief description of cryogenic cloud computing systems is provided in Section 1.3. A quantum computing system and support circuitry are reviewed in Section 1.4. An outline of the book is provided in Section 1.5.

1.1 History of Cryotechnology

Prominent figures in cryogenics include Faraday, Joule, Andrews, Thompson, and Cagniard de la Tour [11, 12]. Joule and Thompson, between 1820 and 1870, established the relationship between the energy of a gas and the temperature and pressure of a gas [13].

The liquefaction of gases, a pivotal advancement in the field of cryogenics, was initially pioneered by Michael Faraday. In 1845, Faraday made a significant contribution to cryogenics by demonstrating the liquefaction of gases, achieving temperatures as low as 173 K [11]. Employing a technique for increasing pressure within a glass tube, Faraday successfully liquefied several gases, such as cyanogen and fluosilicon, under pressure ranging from 50 to 100 atmospheres [14]. Despite these successes, some gases, like hydrogen, oxygen, carbon monoxide, and nitrogen, remained difficult to liquefy at the highest pressures available at the time and were described as permanent gases [15]. Later, both in theory and experiment, it was discovered that these gases had a critical temperature, a temperature above which a gas cannot exist in a liquid state regardless of pressure. Permanent gases had a critical temperature lower than the temperatures achievable at that time.

The critical temperature for carbon dioxide was first identified by the Irish scientist, Thomas Andrews, in 1869 [16]. Subsequent efforts in liquefying gases required a combination of higher pressure and lower temperature. Once a gas is liquefied, the pressure can be reduced, leading to evaporation and a subsequent drop in temperature. This cooler liquid is utilized to liquefy another gas with a lower critical temperature. Numerous gases have been successfully liquefied by sequentially applying this cascadable technique [17].

The liquefaction of the first permanent gas, oxygen, was independently achieved in 1877 by Louis Paul Cailletet in Paris [18] and Raoul-Pierre Pictet in Geneva [19]. Pictet employed a cascadable method using liquid sulfur dioxide and carbon dioxide, while Cailletet utilized cooled liquid ethylene. The liquefaction of nitrogen was achieved in 1883 by Karol Olszewski and Zygmunt Wróblewski in Poland, where they successfully liquefied nitrogen on a large scale [20].

The liquefaction of air, a mixture including permanent gases, was first realized by Carl von Linde in Munich in 1895 [14, 21] using a process that expanded compressed air through a valve, based on the Joule-Thomson effect [13]. This advancement was the result of collaborative efforts by Linde, Hampson, Dewar, and Kamerlingh-Onnes [21]. Shortly thereafter, George Claude liquefied air using a piston engine [14]. The invention of Claude was to use a lubricant which did not freeze at these temperatures – Claude used liquid air itself as a lubricant in the final stage. These methods were later recognized as Linde and Claude cycles [22, 23].

In 1898, British scientist James Dewar liquefied hydrogen using the Linde cycle and was the first to maintain liquid hydrogen using silvered vacuum flasks, now known as dewars [24]. The first successful liquefaction of helium was conducted in 1908 by Dutch scientist Heike Kamerlingh Onnes [25].

The recognition of cryogenic systems gained momentum in the 1960's, leading to a growing need for helium liquefaction, cryogenic insulation containers, advancements in low temperature bearings, and improved lubrication and sealing methods [26]. Developments in cryogenics were often inspired by advances in heat engines. Experiments on an engine by the Stirling brothers in the 1840's led to the development of Gifford-McMahon coolers in 1959 [27–29]. The Stirling cycle was adapted by Kirk in 1862 for an ice making device [30]. This principle was later utilized by Kohler to develop the first commercial air liquefier, known as the "A-machine," within Koninklijke Philips N.V. in 1953, as depicted in Fig. 1.2 [31, 32].

These historical advancements in achieving progressively lower temperatures paved the way for the discovery and exploration of superconductivity, a phenomenon central to the discussions presented in this book. Superconductive circuits, such as single flux quantum circuits [33], can be used in superconductive cloud com-

Fig. 1.2 The A-machine, the pioneering Philips commercial air liquefier [32]. Refrigeration is achieved using a Stirling cycle and hydrogen as a coolant

puting systems and quantum computers, as discussed, respectively, in Sections 1.3 and 1.4.

1.2 Benefits of Cryotechnology

Cryogenic systems, when integrated into electronics, significantly enhance performance. However, assessing the reliability, cost effectiveness, and feasibility is crucial before incorporating cryotechnology into sensors and electronic systems.

Cryogenic operation can boost the performance of IR sensors, electro-optic devices used in commercial, military, industrial applications, and high performance computing systems. Cryogenic technology can enhance communication bandwidths, increase data transfer rates, improve response speeds, enhance reliability, and reduce losses in telecommunication systems [9, 34–36]. The cost effectiveness of cryogenic technology however requires evaluation. Significant performance improvements have been observed in semiconductor devices such as MOSFETs, MESFETs, HEMTs, and CMOS [34], and in optoelectronic and electro-optic components as well as in metal conductors such as aluminum and copper [37]. At extreme cryogenic temperatures, such as liquid helium (4 K), certain materials such as niobium (Nb) can achieve superconductivity [38]. Combining electronics with other cryogenic technologies can justify the cost and complexity of cryogenic cooling, leading to hybrid systems with enhanced performance and reliability. Due to the improved reliability, integrating modular, replaceable electronics can improve system availability, potentially leading to wider acceptance and distribution of cryogenic technologies into a large variety of industrial and commercial applications.

The affordability of cryogenically enhanced systems, including acquisition, lifecycle costs, installation, operation, and maintenance, is vital. Studies suggest that cryogenically cooled systems may be more economically efficient due to enhanced power efficiency as compared to conventional systems operating at room temperature with similar performance [34–36]. Some studies indicate a possible threefold reduction in cost [9]. Furthermore, certain cooling systems such as pulse tube cryogenic coolers [39, 40] exhibit an expected mean time between failure (MTBF) of 10,000 hours [26], highlighting the reliability necessary for critical, difficult to maintain applications such as space missions.

1.3 Cloud Computing Systems

Just as electricity is readily accessible in homes through power outlets, cloud computing provides on-demand computing by connecting users to a cloud-based network. Cloud computing services offer a variety of resources such as storage, processing power, data management, and software applications on a "pay-as-you-go" basis [41]. Typically, cloud computing services are hosted by large scale data centers that house thousands of servers. These data centers run a large variety of software applications and cater to numerous users by virtualizing computing resources. A single physical server in a cloud data center can run multiple operating systems and applications, providing each user with a virtualized computing environment. The data center offers multiple virtual machines (VMs) with diverse operating systems to different users, with a virtual machine monitor managing access for each user, who receives access to an individual VM [42].

Given the stationary nature of cloud computing data centers, incorporating cryogenic cooling can significantly enhance performance and energy efficiency. Despite the lower efficiency of cooling at cryogenic temperatures in small scale applications, a significant increase in efficiency is observed as the scale of the system being cooled increases. For example, at 4 K, the energy efficiency can range from 1% of Carnot efficiency [43] for smaller units to as high as 35% for large scale liquefaction plants [44]. This enhancement in energy efficiency allows large scale cloud computing centers to operate at greater performance while dissipating less power [45].

Note that the power consumption of individual units significantly decreases at cryogenic temperatures. Certain workloads exhibit a hundredfold reduction in power consumption at 77 K as compared to room temperature operation [46]. For instance, a CMOS-based cache operating at 77 K has been reported to consume up to ten times less energy than operating at room temperature for the same workload [47]. Similarly, DRAM at 77 K operates 3.8 times faster while consuming only 9.2% of the power required at room temperature [46]. Furthermore, with the availability of 4 K operation, superconductive logic, such as single flux quantum [33], can be employed to further enhance the energy efficiency of cloud computing systems. Circuits employing passive superconductive interconnects dissipate six orders of magnitude less energy as compared to CMOS interconnects [48].

Cloud computing centers are globally responsible for consuming over 200 TWh of power [49], accounting for more than 2% of the electricity consumption in the

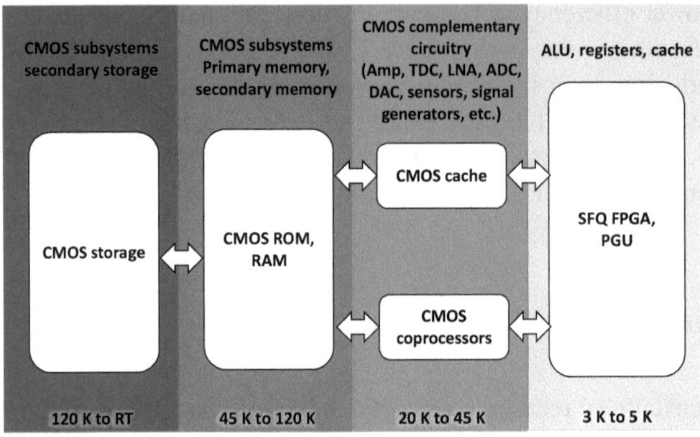

Fig. 1.3 Superconductive cloud computing system. The system consists of several components, such as ROM, RAM, cache, and processors, each operating within a temperature range from 3 K to room temperature

United States [50]. With the growing demand for computational power and the proliferation of data centers, these figures are expected to increase by fourfold by 2030 [49]. Air cooling of data centers consuming over 10 KW is predicted to become unfeasible [51], making cryogenically cooled data centers a potential solution to these escalating energy demands. However, cryogenic cooling for cloud computing centers remains scarce. While liquid nitrogen has been suggested for air cooled data centers [51], bringing the air temperature down to dew point temperatures, these temperatures do not reach cryogenic levels. An example of large scale employment of cryogenic electronics is a superconductive architecture for neural processing units based on superconductive flux quanta [52].

The architecture of a cloud computer includes various components, such as read only memory (ROM), random access memory (RAM), processors, field programmable gate arrays (FPGA), cache, among others, as exemplified in Fig. 1.3. These components fulfill diverse roles such as memory, storage, sensing, and signal generation. Different circuit blocks can be placed at different temperatures, from 3 K to room temperature, depending upon the available cooling power and operational speed of each component. Consequently, optimizing the temperature of each zone within the entire cryogenic system is crucial to enhance overall energy efficiency.

1.4 Quantum Computing Systems

During the 1980's, Richard Feynman identified the inability of classical computers to simulate certain quantum mechanical effects, leading to his speculation concerning the potential efficiency of quantum computing [53]. Classical computers can reduce computational time by using multiple parallel processors. However, to achieve exponential decreases in computational time, the number of processors must increase exponentially. In contrast, quantum computers experience an exponential increase in parallelism with a linear increase in the size of the quantum system, an effect termed quantum parallelism [54, 55]. Quantum computers operate with quantum bits or qubits [56], which differ from classical bits. Unlike classical bits that represent either 0 or 1, qubits can exist in a superposition state, simultaneously embodying both values. Additionally, a set of qubits can represent the superposition of all possible values, meaning a register of n qubits can represent 2^n states [54].

The exponential boost in computational capabilities offered by quantum computers provides a wide range of applications across many fields. These fields include chemistry [57], materials science [58, 59], finance [60], biology [61], artificial intelligence [62], production and logistics [63], and cybersecurity [64]. In chemistry, quantum computers address the "electronic structure problem," characterizing low energy levels in chemical systems [57]. As the number of simulated molecules grows, an exponential number of classical bits is required. A quantum computer only requires a linear growth in the number of qubits. Quantum computers are adept at predicting reaction rates, molecular structures, optical characteristics, and vibrational attributes of molecules [57]. Quantum computers also have potential applications in simulating active sites in enzymes, exploring transition metal nanocatalysts and surface catalysts, and studying light harvesting and vision processes [59].

In the realm of biology, quantum computing offers solutions for challenges such as sequence alignment, computational genetics, and processing X-ray diffraction data [61]. Quantum algorithms can be instrumental in predicting protein folding, calculating ligand binding affinities, and facilitating large scale genomic alignments [61].

The finance sector benefits from quantum computing in areas like option pricing, risk assessment in asset management, and portfolio optimization in investment strategies [60]. Quantum simulations can also enhance portfolio diversification, improve financial forecasting, and refine credit scoring mechanisms in banking [60]. Quantum computing algorithms offer solutions for more effectively tackling these complex financial problems than traditional computational methods.

The core of a quantum computer is the quantum processor, functioning at temperatures of only a few tens of millikelvins [65]. The qubits in these processors can be realized through several methods, including ion traps (ions trapped within a magnetic field), polarized photons, electron or hole spins in semiconductors, transmons, and nitrogen vacancies in diamond lattices [66]. However, qubits typically exhibit short lifespans, only lasting a few nanoseconds [66], requiring a feedback loop to refresh the qubit state [35]. This loop is often provided by a classical controller.

Due to the fragility of the qubit states, direct reading at room temperature is challenging and often infeasible for a large number of qubits [35]. Specialized readout circuitry is therefore required [35, 65]. A quantum computer comprises not just the quantum processor but also a classical controller. This controller manages the qubit control and state readout [35, 66], as depicted in Fig. 1.4. To improve

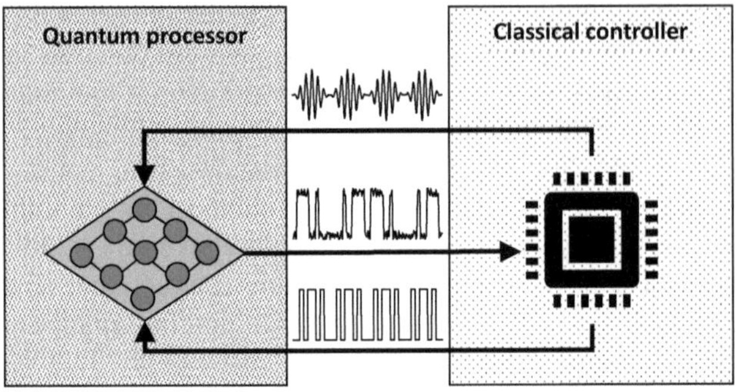

Fig. 1.4 A quantum computer encompassing a quantum processor and a classical control circuit. Both systems operate at cryogenic temperatures

system integration, minimize size, and enhance speed, the controller needs to be located close to the quantum processor within the cryogenic environment [35, 66].

A classical controller in a quantum computing system is responsible for the accurate generation and acquisition of electronic and optical signals, crucial for qubit readout and control [65]. For example, to manipulate the state of certain qubits, such as transmons or spin qubits, microwave signal bursts within the frequency range of four to twelve megahertz are necessary [65]. The readout process for a transmon qubit requires measuring the frequency of a microwave resonator [67], while for a spin qubit, the readout assesses the impedance of a charge detector such as a quantum point contact [68]. Additionally, the controller employs echo techniques and quantum error correction schemes to refresh and correct the qubit states, maintaining the state of a qubit for a longer time [65].

Generating these precise signals for qubit readout and control is currently achieved by CMOS circuits operating at room temperature, where each qubit is individually connected to the CMOS controller [65]. This approach however will become infeasible with millions of qubits in future quantum computers [65]. It has therefore been proposed to employ either CMOS circuits operating at cryogenic temperatures [66] or superconductive logic, such as single flux quantum (SFQ) circuitry [35]. Such a controller would be composed of many components, including voltage, current, and frequency references alongside a digital controller, as illustrated in Fig. 1.5. These components encompass diverse electronic circuits such as SFQ repeaters, phase locked loop oscillators, low noise amplifiers, analog-to-digital converters, and digital-to-analog converters [37]. These circuits are placed in different temperature zones, ranging from a few tens of millikelvins to room temperature [35]. For enhanced speed and power efficiency in quantum computers and cloud computing, the system requires temperature optimization of each component.

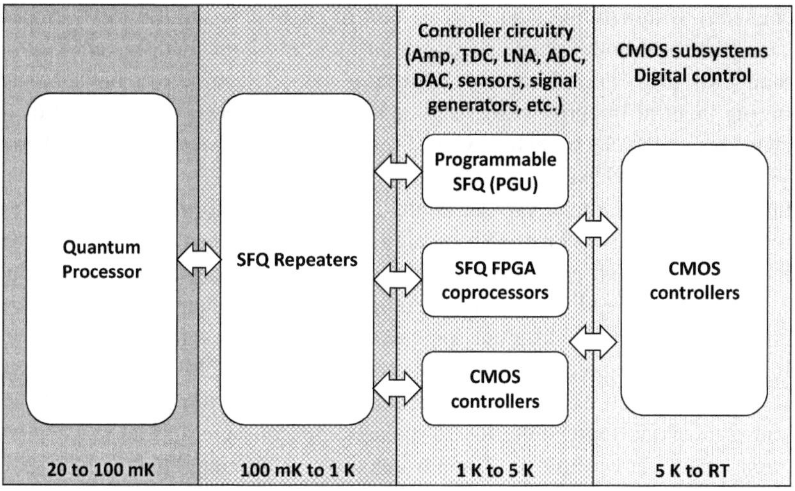

Fig. 1.5 Hybrid quantum-classical computing system. The system consists of different components, including a quantum processor, SFQ repeaters, and several controllers located across a range of temperature zones from a few tens of millikelvins to room temperature

1.5 Outline

The objective of this book is to highlight the advantages of cryogenic computing, particularly focusing on energy efficiency with enhanced performance. Different refrigeration systems and cooling methodologies are reviewed in Chapter 2. Cryogenic cooling systems are categorized into passive and active methods. The fundamental background of the thermodynamics of passive refrigeration systems, which utilize liquid and solid substances as cooling agents, is reviewed. Both liquid and solid passive cooling techniques are discussed, highlighting the different advantages and limitations.

Active refrigeration systems are reviewed in Chapter 2 with an emphasis on assessing the energy efficiency of these systems. Active cryogenic coolers are categorized into two primary types: recuperative and regenerative. Each type of system is reviewed, revealing the operational mechanisms and exploring the merits and disadvantages associated with each system. An understanding of cryogenic cooling and the critical contribution of cryogenic temperatures to efficient and rapid computing is also offered in this chapter.

The exploration of the electronic properties of semiconductor devices, integral to electronic circuits operating at cryogenic temperatures, is the focus of Chapter 3. In this chapter, those factors which determine the electrical conductivity are discussed. Central to this discussion is the role of mobility and density of the charge carriers in determining the electrical conductivity of semiconductor devices. These aspects are influenced by several variables, including energy distribution, effective mass, and band structure, each effect dependent on the temperature. Scattering mechanisms within semiconductors, pivotal in characterizing the carrier mobility, are also discussed in this chapter. Six primary scattering mechanisms are examined, phonon scattering, ionized impurity scattering, velocity saturation, carrier-to-carrier scattering, neutral impurity scattering, and surface scattering. The impact of these mechanisms on the electrical behavior of semiconductor devices at low temperatures is systematically reviewed.

In subsequent chapters, various configurations and optimization techniques for cryogenic electronic systems are presented. In Chapter 4, the utilization of dynamic logic circuits in cloud computing systems at cryogenic temperatures is examined. Due to the widespread use of cloud computing data centers and the availability of cryogenic operation, different logic circuit styles other than traditional static CMOS circuits can be utilized. Unlike traditional static CMOS circuits, dynamic logic circuits, which exhibit certain challenges at room temperature such as the degradation of the logic state due to leakage currents, are appropriate for use in low temperature environments. The advantages of deploying dynamic logic circuits under these conditions are discussed in Chapter 4. A dynamic logic circuit can operate without loosing the logic state at different frequencies depending upon the temperature. A critical temperature and frequency are identified beyond which the use of dynamic logic is more beneficial than static logic. This temperature-frequency breakpoint is also discussed in this chapter. Emphasis is placed on estimating the subthreshold leakage current, a primary leakage current component at cryogenic temperatures. Validation of a leakage current model for dynamic logic is described with a comparison to measured data.

Determining the optimal cryogenic temperature of the supporting circuits within a quantum computing system is described in Chapter 5. This methodology for thermal optimization is based on graph theory. The methodology involves constructing a graph of a system, where each edge represents an individual circuit unit. The edges within a graph are defined by two key parameters: the power consumption and propagation delay associated with each unit. The central objective of the methodology is to identify the most power efficient operating temperature for each unit while maintaining the total delay within a predetermined constraint. An algorithm based on breadth-first search traversal of the graph is proposed, which is integral to the optimization process. Furthermore, the methodology considers the effects of thermal interactions between each unit within a system, as these units are interconnected via cables. A thermal-electrical circuit model is formulated, incorporating the thermal conductivity of these interconnections. A case study based on a quantum computing system is also discussed in Chapter 5. Each component of the readout and control circuitry within the quantum computer is strategically positioned within different temperature zones, extending from 3 K to room temperature, to optimize overall system performance and energy efficiency.

In Chapter 6, exploration of the optimization methodology is further extended to incorporate the mutual dependence of the different temperature zones to enhance the energy efficiency of the overall system. A computing unit can be partitioned in a refrigerator within multiple cooling stages. The same circuitry can be placed in a refrigerator, for example, within the 150 K, 60 K, and 4 K stages, or in a refrigerator within stages at 120 K, 70 K, and 4 K. Differences in performance and power consumption due to these different stage temperatures are discussed in Chapter 6. This enhanced methodology considers the influence of the neighboring stages operating at cryogenic temperatures. The approach represents a transition from individually optimizing each temperature zone to a more comprehensive strategy aimed at lowering overall system-wide energy. The improved methodology is validated by a case study on a cloud computing system, which employs a combination of CMOS and superconductive logic.

Previous approaches to managing this problem, described in Chapters 5 and 6, have relied on a predetermined number of refrigerator stages and fixed unit configurations. For example, the methodology described in Chapter 6 supports several refrigerator stages, but the optimal temperature of each stage is determined for a pre-set number of stages. The units are assigned to each refrigerator chamber in a fixed manner without allowing for grouping of units within one chamber. The methodology proposed in Chapter 7 addresses these limitations by optimizing the number of refrigeration stages based on the specific requirements of the system. This approach allows for more flexible and efficient temperature management across multiple units and temperature zones. Both the number of stages and the configuration of the units within those stages are determined based on performance constraints, maximizing energy efficiency and minimizing power consumption. Furthermore, several optimization techniques, including graph pruning and dynamic programming, are applied to accelerate the search process. As a result, the algorithm determines the optimal number of refrigeration stages, most efficient grouping of units, and corresponding temperature for each unit. The revised methodology is validated by two case studies, a quantum computing system and a cloud computing system.

A summary of the key findings of this book, highlighting the advantages of cryogenic computing systems over traditional computing technologies, is provided in Chapter 8. The ability of cryogenic computing systems to overcome limitations faced by conventional systems is presented. The impact of this book on the field of computing systems is also discussed.

Chapter 2
Cryogenic Systems for Electronic Applications

A variety of methods exist for achieving cryogenic temperatures for electronic applications. The choice of cooling method requires making a decision based on tradeoffs among cost, physical space, weight, reliability, vibration, and electromagnetic noise. Refrigeration systems for achieving cryogenic temperatures fall into two main categories: passive and active. Passive refrigeration systems employ a simple design based on dewar storage, a vacuum flask containing a liquid or solid coolant termed a cryogen. These refrigeration systems function by immersing the object to be cooled into a container filled with a liquid or solid cryogen. Despite offering some advantages, such as stability and lower noise, passive systems require regular replenishment of the cryogen. In contrast, active refrigeration systems are closed loop systems, eliminating the need for frequent refilling of the cryogen. These systems include motors and compressors and operate based on engineering principles such as Stirling, pulse tube, Gifford-McMahon, Joule-Thomson, Claude, and Brayton cycles. Different types of passive and active refrigeration systems, principles of operation, and system efficiency are discussed in this chapter.

2.1 Passive Refrigeration Systems

Passive refrigeration systems, utilizing stored cryogens in liquid and solid forms, have proven to be highly reliable for achieving cryogenic temperatures [69]. These systems operate based on the boiling or sublimation of cryogens such as liquid helium, liquid nitrogen, or solid ammonia [23]. Modification of the system pressure supports the control of the boiling and sublimation points to achieve a desired temperature. Passive systems offer advantages such as a stable temperature, minimal noise from vibrational and electromagnetic effects, and low power. A significant drawback, however, is the continual need to replenish the cryogen. Additionally,

N. Zhuldassov, E. G. Friedman, *Cryogenic Microelectronic Systems for Ultra-Low Energy and Enhanced Performance*, https://doi.org/10.1007/978-3-031-90754-8_2

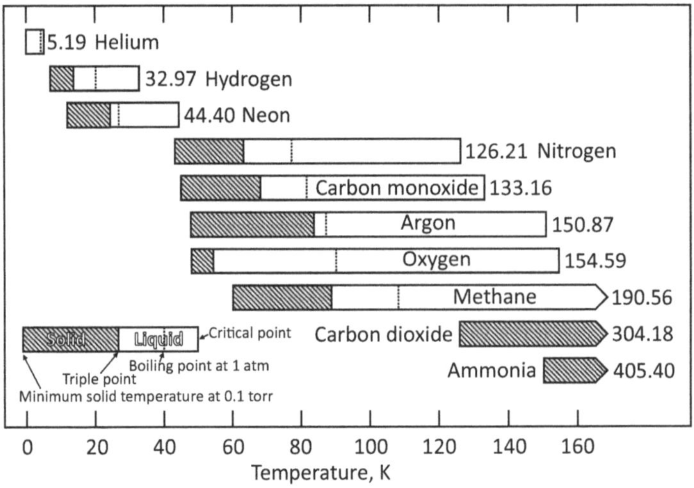

Fig. 2.1 Commonly used coolants and relevant operating temperatures [69–72]

these systems often have limited temperature control capabilities as compared to active systems.

Liquid cryogenic systems are categorized into three types: subcritical systems with standard liquid and vapor phases, densified systems featuring low pressure liquid and vapor, and supercritical systems characterized by a single phase [69]. Subcritical systems, such as helium at low pressure, are commonly used for temperatures below 4 K. Solid cryogens are typically used at the triple point where direct sublimation from solid to vapor occurs. Solid cryogenic systems provide more stable temperature control as compared to liquid systems. Additionally, these systems offer higher mass and volume density and greater heat capacity. Examples of commonly used coolants along with operating temperatures are illustrated in Fig. 2.1 [69–72]. As noted in the figure, most of the coolants can be used in all three phases, subcritical, densified, and supercritical. Thermodynamic principles underlying passive refrigeration systems and the functionality and applications of these systems, including both liquid and solid cryogenic systems, are reviewed in this section.

2.1.1 Thermodynamic Principles of Passive Cooling Systems

To describe the thermodynamic behavior of applying a cryogen to a cryogenic system, it is essential to review the concepts of heat transfer and entropy. Entropy in a system is determined by the quantity of transferred heat and temperature [73],

$$\Delta S = \frac{\Delta Q}{T}. \tag{2.1}$$

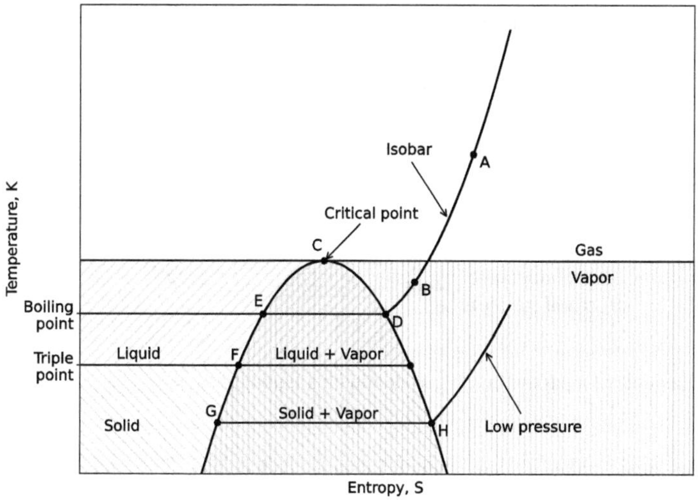

Fig. 2.2 Temperature-entropy diagram of liquid cryogen

An ideal depiction of the temperature-entropy relationship for a liquid cryogen is illustrated in Fig. 2.2. The point C in the diagram is known as the critical point of a material. The cryogen cannot exist in a liquid state above the critical point, and above C exists solely as a homogeneous gas.

The cooling process of a gas under constant pressure is traced along the path ABDE. For example, water at atmospheric pressure cools from a gaseous state to vapor, becoming 100% saturated vapor at point D. At this temperature, water remains at a constant temperature of 100 °C as water transitions into a 100% saturated liquid at point E. Subsequent reductions in temperature lead to liquid freezing into a solid at point F, the triple point. Here, at 0°C for water, the substance simultaneously exists in solid, liquid, and gaseous states. Any further temperature reduction, as represented by point G, requires the addition of heat at a specific pressure, causing the solid to sublimate. Sublimation is a process where a solid material makes a direct transition from a solid state into a gaseous state. This transition is represented by the line GH in the diagram, which represents the relationship among temperature, pressure, and phase changes in cryogenic substances.

2.1.2 Liquid Cryogenic Systems

A passive storage system with a liquid cryogen is illustrated in Fig. 2.3. The system includes nested storage tanks separated by insulation. The innermost tank contains the liquid cryogen. These storage tanks are interconnected through struts

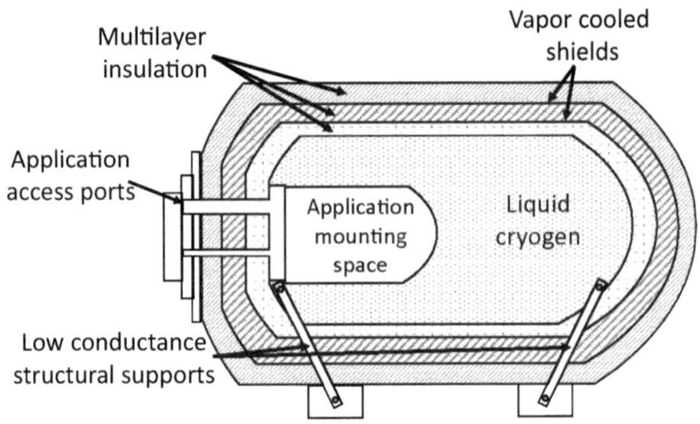

Fig. 2.3 Example of a passive cooling system with a liquid cryogen

and cables designed with low thermal conductivity [69]. The space between these tanks is packed with multilayer insulation. Additionally, the evaporating cryogen is further exploited by incorporating vapor cooled shields within the tank to enhance efficiency [74].

When liquid helium is utilized to achieve 2 K, specific considerations for effective cooling need to be considered. At this temperature, helium transitions into a superfluid state, exhibiting unique characteristics [75]. In this state, helium has zero viscosity and exhibits infinite thermal conductivity. These properties enable the helium to flow without friction over any surface. Superfluid helium is able to penetrate through the pores of a container [76, 77]. Specially designed porous plugs are employed in conjunction with superfluid helium to address this issue [76]. These porous plugs contain micrometer sized pores which allow controlled separation of the fluid from the vapor, preventing premature escape of the liquid helium [76, 78].

2.1.3 Solid Cryogenic Systems

Passive cooling systems can effectively utilize solid cryogens. In these systems, the cryogen is employed in a solid form, operating in the region below the triple point, as indicated in Fig. 2.2. Under pressure, the solid transitions directly into vapor, absorbing heat in the process. The use of solid cryogens offers several benefits, including reduced volume and mass of the cooling system due to a higher heat absorption density, both per volume and mass [23]. Furthermore, solid cryogens can accommodate a broader range of temperatures as compared to liquid cryogens. Additionally, solid cryogenic systems may be preferable due to certain challenges associated with liquid cryogens, such as phase separation and operational noise. One primary limitation, however, is the relatively small variety of available solid cryogens. Additionally, systems using solid cryogens typically demand more

Table 2.1 Thermal properties of solid cryogens [69, 79]

Solid cryogen	Heat of sublimation, J/g	Solid density at melting point, kg/m^3	Temperature range, K	
			0.1 Torr	Triple point
Ammonia (NH_3)	1719	822	150	195
Carbon dioxide (CO_2)	574	1562	125	216
Methane (CH_4)	569	498	60	90
Oxygen (O_2)	227	1302	48	55
Argon (A)	186	1714	48	84
Carbon monoxide (CO)	293	929	46	68
Nitrogen (N_2)	225	1022	43	63
Neon (Ne)	106	1439	14	25
Hydrogen (H_2)	508	80.4	8	14

frequent maintenance and manpower. The heat properties of some solid cryogens are listed in Table 2.1 [69, 79].

As noted in Table 2.1, the temperature achievable by a specific cryogen is contingent upon the minimum pressure that can be maintained. The highest temperature a cryogen can provide is set by the triple point temperature. The temperature can be modulated between these points depending upon the available pressure.

Dewar systems for solid cryogens are similar to dewars for liquid cryogens, as depicted in Fig. 2.3, featuring nested storage tanks. There are, however, two primary components distinctive to solid cryogen dewars: a thermally conductive metallic matrix and freezing coils [69]. The metallic matrix maintains a thermal link between the cryogen and the load. The thermal link maintains a consistent thermal connection during sublimation of the cryogen. Freezing coils, cooled by an external coolant, solidify the cryogen and maintain a controlled rate of sublimation. Additionally, freezing coils are also used to precool the cryogen to a temperature below the freezing temperature of the particular cryogen.

2.2 Active Refrigeration Systems

The Helium Conservation Act of 1925 [80], enacted by the US government, imposed regulations on helium production and storage to conserve helium. Despite subsequent relaxations of this act, helium remains scarce and expensive [44]. The scarcity, coupled with the inconvenience of passive systems for certain autonomous applications, has led to the adoption of active refrigeration systems. Different refrigeration techniques are employed in these active systems, such as Stirling, Brayton, Gifford-McMahon (GM), Claude, and Joule-Thomson (JT) cycles. The principles of operation of these cryocoolers and techniques for measuring the efficiency of these cryocoolers are reviewed in this section.

2.2.1 Assessing Efficiency of Cryocoolers

The efficiency of cooling in cryocoolers is commonly measured using the coefficient of performance (COP), also known as Carnot efficiency [43]. It is defined as the ratio of the cooling power to the input power,

$$COP_{Carnot} \equiv \frac{Cooling\ power}{Input\ power}. \tag{2.2}$$

This expression describes the ideal efficiency based on the Carnot cycle [43], which is depicted in Fig. 2.4. The Carnot cycle is comprised of four phases: refrigeration (line 4 to 1), heat rejection (line 2 to 3), gas compression (line 1 to 2), and gas expansion (line 3 to 4). During the heat rejection phase, the heat of the cryogen is absorbed by the environment. Note that the heat rejection and heat absorption phases are isothermal (at constant temperature), and the gas compression and expansion phases are isentropic (constant entropy), i.e., without heat transfer.

From the Carnot refrigeration cycle, the Carnot efficiency is the ratio of the absorbed heat to the input power. The input power is the heat rejected minus the cooling power. Consequently, the Carnot efficiency is solely a function of temperature,

$$COP_{Carnot} = \eta_C = \frac{Cooling\ power}{Input\ power} = \frac{T_c}{T_h - T_c}, \tag{2.3}$$

where T_c is the temperature of cooling, and T_h is the ambient temperature of the environment.

Practically, however, cryocoolers are not ideal and the efficiency is assessed as a per cent of the Carnot efficiency. This efficiency is referred to as the second law

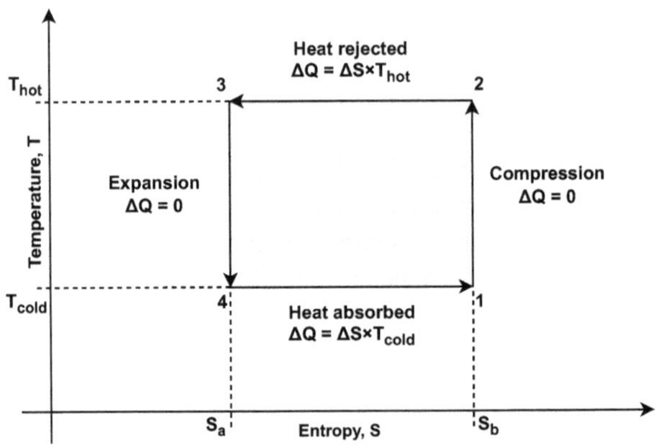

Fig. 2.4 Carnot refrigeration cycle

efficiency [44]. The second law efficiency is

$$\eta_{II} = \frac{\eta}{\eta_C} \times 100\%, \tag{2.4}$$

where η is the efficiency of the cryocooler.

The efficiency of cryocoolers range from 15% to 25% of COP in 80 K cryocoolers [44]. For 4 K cryocoolers, the efficiency ranges from 1% of COP in smaller units to 35% of COP in large liquefaction plants [44].

The overall efficiency of a cryocooler is primarily influenced by the thermodynamic efficiency of the compressor/expander and the efficiency of the compressor drive motor [69]. The compressor/expander thermodynamic efficiency is the efficiency of the refrigeration cycle and heat transfer process, while the compressor drive motor efficiency is the electrical efficiency of the compressor motor. Dividing the cryocooler efficiency into these two components allows for targeted improvements in certain parameters. The thermodynamic efficiency is the efficiency of expanding and compressing the gas [69],

$$\eta_{Thermodynamic} = \frac{Cooling\ power}{PV\ work} \approx \frac{Cooling\ power}{input\ electrical\ power - I^2R}. \tag{2.5}$$

Here, PV work indicates the pressure-volume work exerted on the gas, calculated by deducting the I^2R losses of the motor from the total input power. Other compressor losses, such as friction and eddy current losses, are relatively minor as compared to I^2R losses.

2.2.2 Cryocooler Classifications

Cryocoolers function by expanding a gas under high pressure, absorbing the heat. The classification of cryocoolers is based on several factors, including the method of compression and expansion, the pressure to expansion ratio, and the technique for heat rejection. A fundamental distinction between cryocoolers is based on the refrigerant flow: DC systems with continuous flow of the cryogen and AC systems with oscillating flow of the cryogen.

In DC systems, the refrigerant flows steadily in one direction, facilitated by different inlet and outlet valves, similar to DC electrical systems. These systems are also known as recuperative systems [69]. A key component of these systems is a heat exchanger (or recuperator), which transfers heat between two streams of gas or liquid refrigerant flowing in opposite directions. Recuperative compressors typically employ oil lubrication and require mechanisms to extract oil from the cold end of a cooler, preventing oil from freezing. The three most common recuperative cryocoolers are Joule–Thomson [13, 23], Brayton [23], and Claude cycle [22, 23] cryocoolers, illustrated in Fig. 2.5. The Claude cycle combines elements of both

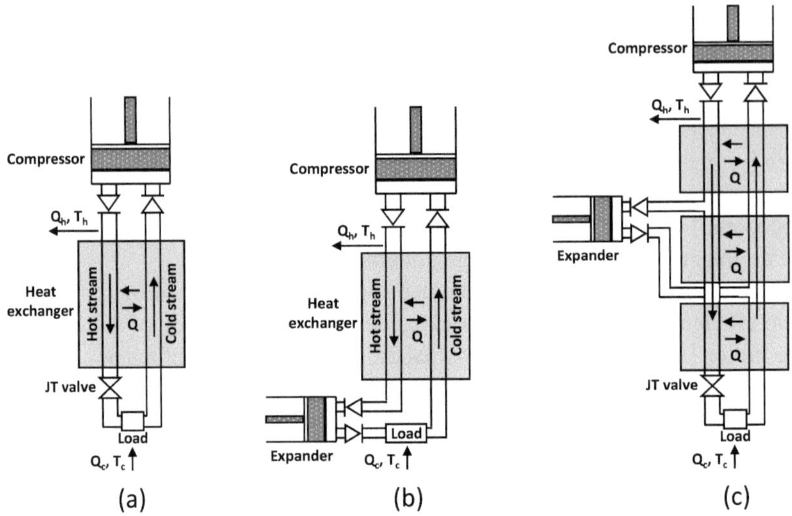

Fig. 2.5 Most common recuperative cryocooler systems, (a) Joule-Thomson cryocooler, (b) Brayton cryocooler, and (c) Claude cryocooler. Q_h and T_h are, respectively, heat rejected to the environment and the ambient temperature of the environment. Q_c and T_c are, respectively, the heat absorbed from the cooled load and the temperature of the cold end

the Brayton and JT cycles, using the JT cycle to liquefy the working fluid during expansion. These cryocoolers are employed in a variety of applications, such as large helium liquefaction plants, cooling superconducting magnets, and RF cavities in particle accelerators [81]. A key advantage of recuperative systems is the ability to position the compressor farther from the cooling end, thereby reducing noise and vibrational effects.

In AC cryogenic systems, the flow and pressure of the refrigerant oscillate, similar to AC electrical currents. These systems are commonly known as regenerative systems, characterized by the use of a regenerator for heat exchange within the alternating refrigerant stream [69]. Stirling [23, 30], Gifford-McMahon [82, 83], and pulse tube cryocoolers [39, 40] are the most common types of regenerative cryocoolers [44], which are depicted in Fig. 2.6.

In these systems, the pressure typically ranges from 1.5 MPa to 3 MPa, with oscillation amplitudes of about 10 to 20% [44]. The operational frequency for regenerative systems is generally around 1 Hz for pulse tube and GM cryocoolers, and can reach up to 60 Hz in Stirling and some pulse tube cryocoolers [44]. Stirling cycle cryocoolers employ a reciprocating compressor, a compressor without valves, to generate an oscillating pressure. Alternatively, GM cryocoolers utilize a standard compressor with a set of valves that alternate between different pressure levels. Due to these valves, the compressors in GM coolers can be positioned farther from the cooling end, similar to recuperative systems. In contrast, Stirling cryocoolers typically require the compressor to be placed closer to the cooling end.

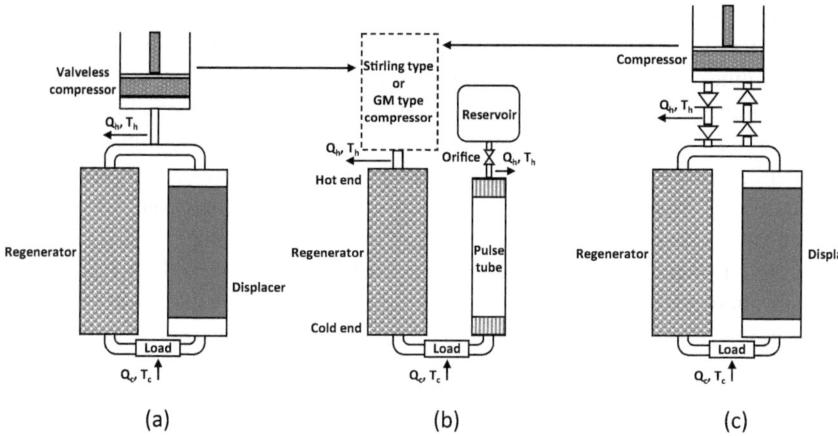

Fig. 2.6 Most common regenerative cryocooler systems, (a) Stirling cryocooler, (b) pulse tube cryocooler, and (c) Gifford-McMahon cryocooler

2.2.3 Recuperative Systems

Some examples of recuperative cryocoolers are depicted in Fig. 2.5. The cryogen in these cryocoolers flow in one direction under constant pressure. The three most prevalent recuperative systems with DC flow of cryogen, Joule–Thomson, Brayton, and Claude cryocoolers, are reviewed in this section.

Joule-Thomson Cryocoolers

In Joule-Thomson cryocoolers, as depicted in Fig. 2.5a, cooling is achieved through the passive expansion of a gas. A cryogenic gas under high pressure expands as the gas flows through a JT valve, which can be a specialized valve, porous plug, or capillary [44, 84]. The expansion process does not require production of any work, thus occurring under a constant enthalpy, the sum of the internal energy of the gas and product of pressure and volume (PV work) [85]. Heat absorption is performed by the cooled or liquefied gas. After heat absorption, the gas cycles back to the heat exchanger to precool the incoming high pressure gas. Although in an ideal gas the enthalpy is independent of pressure, the enthalpy changes with pressure in real gases (non-ideal). In real gases, the product PV is not constant, leading to a change in the temperature and internal energy as opposed to in ideal gases. Joule-Thomson coolers require a pressure of 20 MPa or more [86], which includes the high performance compressors that often exhibit a limited lifespan [87].

The advantages of Joule-Thomson cryocoolers include the absence of moving parts in the cold section, eliminating the need for lubrication and allowing miniaturization. Furthermore, cooling in a JT refrigerator is a rapid process [87], finding applications in military systems such as missile guidance, where nitrogen or argon is at 45 MPa [44]. The cryogen in these systems is not reused, providing cooling for several minutes until the cryogen is fully depleted [44].

The requirement for a small orifice however poses a threat of clogging. Furthermore, the high pressure compressors used in closed cycle systems is another disadvantage. In addition to shorter lifetimes, these high pressure compressors are often inefficient [87]. Below 70 K, a JT cooler requires a second compressor and a cryogen such as neon or hydrogen for precooling the second stage [44].

Recent advancements in Joule-Thomson coolers include mixed refrigerant cascades, as opposed to the use of pure gases [17, 84, 86]. Cryogens with higher boiling temperatures are mixed with lower boiling temperature gases to produce a greater change in enthalpy with pressure, which causes the cryogen to behave as a real gas over a wide temperature range [88]. For instance, methane, ethane, or propane can be mixed with nitrogen [17], allowing operation at a lower pressure (around 2.5 MPa) [44]. These low pressure JT systems have extended lifespans of up to ten years at 80 K [44]. The application of helium operated Joule-Thomson coolers include cooling the Mid-Infrared Instrument on the James Webb Telescope to 7 K [4] and single photon detectors to 2 K [89].

Brayton Cryocoolers

Brayton cryocoolers, as illustrated in Fig. 2.5b, employ either a piston or a turbine with aerodynamic bearings for the gas expansion process. The work produced by the piston in an ideal gas is equal to the absorbed heat. The gas, once cooled through expansion, absorbs heat from the load and precools the incoming high pressure gas in a heat exchanger. This process is generally more efficient than in Joule-Thomson coolers and requires a lower pressure. An additional advantage of turbines with air bearings is reduced vibrations. However, manufacturing miniature pistons or turbines poses significant challenges [44].

In space applications, turbines as small as 6 millimeters in diameter, rotating at speeds between 2000 to 5000 Hz, are utilized [90]. A centrifugal compressor with a compression ratio of 1.6 and 100 kPa at the lower pressure side is used to cool infrared sensors [90]. The Hubble telescope employs a Brayton cycle cryocooler offering 8 W of cooling power at 70 K with an efficiency of 8% of the Carnot cycle [91]. Typically, neon is used as the cryogen for temperatures above 35 K, while helium is preferred for lower temperatures [44]. It is important, however, to note that the efficiency of Brayton cryocoolers is generally lower than Stirling cryocoolers and some pulse tube cryocoolers [44].

Claude Cryocoolers

The Claude cycle cryocooler, as depicted in Fig. 2.5c, requires precooling a high pressure gas before expansion. The Claude cycle incorporates multiple stages, each employing either Joule-Thomson or Brayton cycles. Due to the precooling process in the final stage, the gas often liquefies, making the use of pistons for expansion infeasible. Consequently, the JT cycle is typically adopted for the final stage. In the Claude cycle, the initial stage often consists of a Brayton cycle cryocooler which uses reciprocating engines for gas cooling via expansion. This cooled gas is circulated back to the heat exchanger to precool the incoming gas under high pressure, as illustrated in Fig. 2.5c. Subsequently, the gas enters the second stage of the cryocooler where a JT cycle is employed. The gas expands through the JT valves, which cools or condenses the gas into a liquid. This cooled or liquefied gas absorbs heat from the load and flows back to the heat exchanger for a second precooling phase.

Claude cycle cryocoolers are predominantly used in liquefaction plants to liquefy air, hydrogen, helium, and nitrogen [44]. The first commercial helium liquefier, introduced in 1947, incorporated three stages: the first two stages use expansion engines (Brayton cycle) and the final stage utilizes JT expansion valves (JT cycle) [92]. The liquefaction rates of these cryocoolers range from about 190 liters per day to approximately 400 liters per day, which can be doubled if precooling with liquid nitrogen is employed [44]. Helium liquefiers typically include helium recovery and purification systems to recapture and recycle boiled off helium.

The efficiency of Claude cycle cryocoolers without nitrogen precooling generally ranges from 5% to 8% of the Carnot efficiency, with larger systems achieving up to 12% [44]. The largest helium liquefaction system in the world, installed at the Large Hadron Collider, is comprised of eight cryocoolers operating at 4.5 K with 18 kW of cooling power each and eight cryocoolers at 1.8 K with 2.4 kW each [7]. These cryocoolers achieve a second law efficiency of up to 29% of the Carnot efficiency [7]. To achieve this efficiency, eight to ten stages are employed within the system.

2.2.4 Regenerative Systems

Regenerative cryocooler systems, as illustrated in Fig. 2.6, differ from recuperative systems by employing an oscillating flow of cryogen, which results in fluctuating pressure levels. This oscillating flow is in contrast to the constant flow observed in recuperative systems. Helium is a common cryogen used in regenerative systems [44].

The heat exchanger in these systems is called a regenerator. The regenerator absorbs heat from the cryogen as the heat flows from the hot end of the cryocooler. Conversely, during the return cycle, the cryogen moves back from the cold end, cooling the regenerator to the initial temperature. For effective heat absorption, regenerators often incorporate meshed screens or packed spheres [44, 69], designed to maximize surface area and enhance the efficiency of the heat transfer process.

Stirling cycle regenerative cryocoolers uniquely utilize valveless compressors to create alternating pressure, unlike Gifford-McMahon (GM) cycle systems that employ traditional compressors in conjunction with valves. The use of constant pressure compressors in GM cryocoolers renders these cryocoolers more cost effective, albeit with a compromise in efficiency [44]. Pulse tube cryocoolers represent a versatile category, which operate on either the Stirling or GM cycle to generate an oscillating pressure. The characteristics and functionality of these three types of regenerative cryocoolers are reviewed in this section.

Stirling Cryocoolers

The Stirling cycle, initially conceptualized by Robert Stirling in 1815, found its first application in cooling technologies in 1867 [30]. The principle was later employed in a commercial cooler for air liquefaction in 1953 [32]. The Stirling cryocooler, as depicted in Fig. 2.6a, does not facilitate rapid heat absorption as in JT cryocoolers but offers enhanced efficiency. Stirling coolers typically provide a cooling power ranging from 0.15 to 1.75 watts [44].

An oscillating pressure within a closed confinement leads to fluctuating temperatures. This temperature oscillation is utilized in refrigeration by employing a displacer to separate the phases of oscillation in the Stirling cryocoolers. The displacer, a movable component, illustrated in Fig. 2.6a, transfers gas between the hot and cold ends. When the displacer moves downwards, gas is directed to the warmer end through the regenerator. Subsequently, a piston compresses the cryogen, raising the temperature. After the heat is dissipated into the environment, the displacer moves upwards, channeling the cryogen to the colder end via the regenerator. The piston expands the cryogen, cooling the load in the process. In Stirling cryocoolers, the regenerator is typically integrated within the displacer [44]. The piston, which generates pressure, and the displacer move in a sinusoidal pattern, phased 90 degrees apart [93]. The phase difference ensures synchronization of the piston movement with the mass flow of the cryogen through the regenerator. In an ideal scenario, where compression and expansion are isothermal, the efficiency of the Stirling cycle is the same as the Carnot efficiency.

The Mean Time To Failure (MTTF) of standard Stirling cryocoolers often exceeds 4,000 hours, with some models achieving over 10,000 hours [44]. Stirling cryocoolers designated for space applications have demonstrated MTTF rates up to ten years [44].

Gifford-McMahon Cryocoolers

Gifford-McMahon cryocoolers, as displayed in Fig. 2.6c, diverge from Stirling cryocoolers primarily by the compression/expansion method. A conventional compressor, which operates at a consistent pressure, is used in GM coolers, combined with a range of valve systems to generate oscillating pressure [23, 94]. This design leads to two principal benefits: reduced noise and higher cost efficiency [23, 94]. The ability to use a standard commercial compressor, which can be found in refrigerators or air conditioning units, significantly reduces the cost of the cooler. Additionally, the constant pressure allows the use of flexible hoses, enabling the compressor to be positioned several meters from the cooling end, which is advantageous in terms of noise reduction and system layout flexibility [44]. This system, however, requires oil for lubrication, necessitating oil removal equipment to avoid freezing of the oil.

GM cryocoolers are frequently utilized in magnetic resonance imaging devices to eliminate the need for continuous helium replenishment [44]. These coolers can achieve a temperature of 30 K in single stage configurations [44]. A significant advancement occurred in 1990 when Er_3Ni spheres [95], rather than traditional materials, were used within the regenerator, allowing, for the first time, two stage GM cryocoolers to reach 4.5 K. Commercial GM cryocoolers can provide cooling capacities between 0.1 and 2 watts at 4.2 K.

Pulse Tube Cryocoolers

The pulse tube cryocooler, first developed in 1963 by Gifford and Longsworth [39], is shown in Fig. 2.6b. As noted in the figure, a pulse tube cryocooler does not have a displacer as opposed to Stirling and GM cryocoolers. The absence of a displacer provides several benefits to the pulse tube cryocooler, such as diminished vibration and electromagnetic noise alongside enhanced reliability and longer lifetime. The displacement of gas, typically achieved by a displacer in other coolers, is accomplished in pulse tube cryocoolers through a combination of a specialized orifice and a reservoir. The volume of the reservoir is selected to minimize any oscillations in pressure [44]. The dynamic pressure within the pulse tube creates a drop in pressure across the orifice, which propels gas through the orifice. The gas movement at the cold end is similar to systems with a displacer, achieved via gas flow through the orifice. Unlike systems with a displacer, where gas at the cold end is in phase with the regenerator phase, gas oscillation in this system is in phase with the oscillation in the orifice. The separation of cold and hot phases is achieved by the oscillation in the orifice, replacing the role of a displacer. Compression and expansion inside the pulse tube should ideally be an adiabatic process, thereby limiting the minimum diameter of the pulse tube.

Pulse tube operation involves four phases. Initially (1), the piston compresses the gas in the tube, followed by (2), the forced movement of the gas into the reservoir via the orifice due to elevated pressure. This heated gas dissipates the energy in the heat exchanger when flowing through the orifice. Upon pressure equalization, gas flow through the orifice stops. Subsequently (3), the piston adiabatically expands the gas. In the final phase (4), cooled gas returns from the reservoir through the orifice, absorbing heat from the load. The movement of gas stops once the pressure is equalized and the process repeats [96].

The size of the pulse tube is dictated by the need to insulate the cold and hot gas sections within the pulse tube. Only a portion of the gas flows to the reservoir from the tube. The gas at the middle section of the pulse tube acts as an insulator, effectively replacing the displacer. To improve this insulation, any turbulence

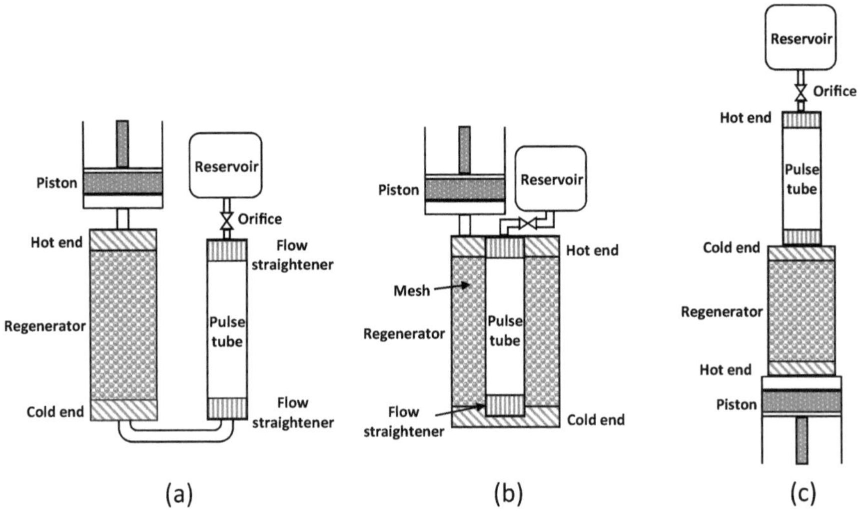

Fig. 2.7 Three common geometries of pulse tube cryocooler: (a) u-tube, (b) coaxial, and (c) inline

within the pulse tube needs to be minimized. To achieve minimal turbulence, flow straightening screens are employed at both ends of the tube [44, 96].

The earliest pulse tube cryocoolers did not employ an orifice and achieved temperatures of 124 K [39]. The introduction of an orifice in 1984 facilitated a temperature reduction to 105 K [97], with further advancements in 1986 lowering the temperature to 60 K [98]. By the 2000's, single stage pulse tube cryocoolers were reaching 20 K, with two stage cryocoolers attaining temperatures as low as 2 K [99].

Different configurations of pulse tube cryocoolers offer distinct advantages and drawbacks. The most commonly used types include u-tube, coaxial, and inline cryocoolers, as displayed in Fig. 2.7 [96]. The most efficient configuration is the inline geometry. In u-tube and coaxial geometries, the direction of the flow of gas changes, which requires extra space and reduces the efficiency. In contrast, the inline configuration maintains a direct flow path, resulting in reduced turbulence as compared to the other two types (u-tube and coaxial).

The coaxial geometry is more compact as compared to other configurations. The disadvantage of this geometry is lower efficiency due to heat exchange between the regenerator and the pulse tube. However, advancements have led to the development of a coaxial pulse tube cryocooler with 20% of Carnot efficiency (second law efficiency) at 90 K, capable of providing 18.8 W of cooling power with an input of 222 W. The cryocooler was designed for NASA in 2000 to liquefy oxygen [100]. In 2015, a hybrid Stirling pulse tube cryocooler was developed, achieving an efficiency of 24.2% of Carnot at 80 K [101].

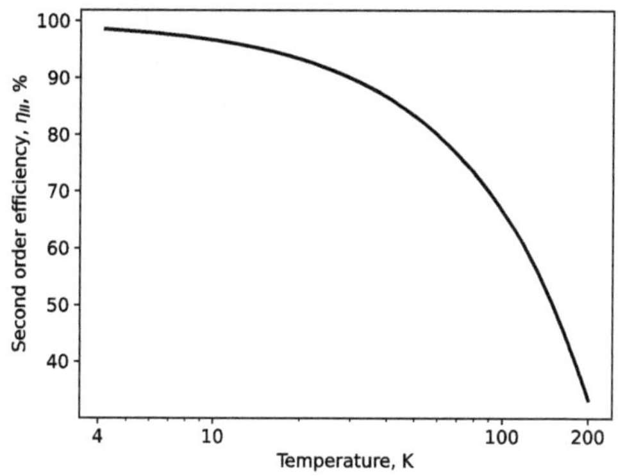

Fig. 2.8 Second order efficiency of an ideal pulse tube cryocooler. The ambient temperature is set at 300 K

Despite these advantages, pulse tube cryocoolers face challenges in scalability and efficiency. Small scale cryocoolers with an input power below 10 W are difficult to develop due to constraints on the minimum size of the pulse tube [44]. Furthermore, the absence of a mechanical displacer prevents the efficiency of a cryocooler from reaching 100% of Carnot even in ideal conditions. The potential energy of a forced displacer movement produces work on both sides of a displacer, which reduces the work done by the compressor. In a pulse tube cryocooler, this potential energy is dissipated in the orifice [96]. The COP of a pulse tube cryocooler is therefore limited to [102]

$$COP_{pulse\,tube} = \eta_{pulse\,tube} = \frac{T_c}{T_h}. \tag{2.6}$$

The second order efficiency of an ideal pulse tube cryocooler operating at an ambient temperature of 300 K is depicted in Fig. 2.8. As evident from the plot, note that at 77 K the maximum efficiency is limited to 74% of Carnot for an ideal system.

2.3 Conclusions

Different means to achieve cryogenic temperatures, categorized into passive and active refrigeration systems, are reviewed in this chapter. Passive systems, typically based on dewar technology, require regular refilling with cryogenic liquids or solids. The benefits of passive systems include low noise levels, minimal vibration, high reliability, and affordable initial costs. Conversely, active refrigeration systems are essential for achieving lower operational costs, autonomous functioning, and the production of gases used in passive systems. Active systems are bifurcated into recuperative and regenerative types, distinguished by operation under, respectively, constant and oscillating pressures. Selecting an appropriate refrigeration method requires careful consideration of several factors, including cost, efficiency, size, and noise level. Assessing the efficiency of different refrigeration methods at different temperatures is crucial when optimizing the thermal behavior of the entire system [36, 36, 37]. This issue is pivotal to ensure optimal performance and cost effectiveness in applications requiring cryogenic operation.

Chapter 3
Semiconductor Behavior at Cryogenic Temperatures

Understanding changes in the behavior of semiconductor conductivity as the temperature decreases requires knowledge of the carrier mobility and the density of the free carriers at cryogenic temperatures. Carrier mobility is influenced by six primary scattering mechanisms: phonon scattering, ionized impurity scattering, velocity saturation, carrier-to-carrier scattering, neutral impurity scattering, and surface scattering. To fully comprehend carrier behavior in intrinsic and doped silicon, it is important to examine how the electron and hole populations, along with the Fermi level, change with decreasing temperature. This discussion also leads to an analysis of the band structure of silicon and variations across different temperatures. Additionally, it is necessary to discuss how the effective mass in different bands changes with temperature. These elements are crucial for understanding the electronic structure and energy distribution within semiconductors under cryogenic conditions. An overview of these topics is provided in this chapter.

3.1 Energy Distribution and Charge Carriers

The electronic characteristics and configuration of semiconductors have been extensively examined since diodes became commonplace during the 1960's [103]. Research exists to forecast the electronic, optical, magnetic, thermal, and mechanical attributes of common semiconductors operating at cryogenic temperatures.

To understand these semiconductor characteristics, an in-depth exploration of the concepts of band theory and quantum mechanics is essential. Furthermore, understanding these concepts extends beyond a theoretical analysis to enable the effective design of integrated circuits operating at cryogenic temperatures.

The quantum mechanical features of semiconductors, including the effective mass m^*, intrinsic concentration n_i, effective density of states U_c (conduction band) and U_v (valence band), absorption coefficients α_n and α_p, Fermi level E_f, along

N. Zhuldassov, E. G. Friedman, *Cryogenic Microelectronic Systems for Ultra-Low Energy and Enhanced Performance*, https://doi.org/10.1007/978-3-031-90754-8_3

with electron and hole relaxation times, respectively, τ_n and τ_p, are instrumental in characterizing the electronic attributes of semiconductors operating at cryogenic temperatures. Moreover, these parameters are intricately interconnected, exhibiting complex interactions at different temperatures, complicating the prediction of optical, thermal, and electronic behaviors. Electron scattering is critical in analyzing electronic properties at reduced temperatures [104]. Scattering within the semiconductor lattice affects the carrier mobility μ, which, combined with the free carrier density n or p, determines the conductivity of semiconductors. A lower temperature leads to incomplete ionization of the dopant atoms. Due to this effect, if the doping concentration is less than 1×10^{18} atoms per cm^3, shallow impurity traps emerge when temperatures are below approximately 100 K [104]. These traps, once occupied by free carriers, diminish the free carrier density. With a further decrease in temperature below 30 K, the carriers are absent from the conduction bands, leading to a phenomenon known as carrier freezeout [105]. The onset of carrier freezeout precipitates a steep decline in semiconductor conductivity as the temperature drops [106].

Moreover, the electrical conductivity of semiconductors at low temperatures is influenced by electric fields [107–110]. These fields can induce a hot electron effect, which impairs MOSFET performance and long term reliability [111]. Furthermore, these fields lead to velocity overshoot, resulting in swifter than anticipated device switching [112], and can cause ballistic conduction, particularly in short-channel devices [113].

3.1.1 Energy Bands

Understanding the change in charge carrier density and mobility with temperature requires an examination of the energy bands and how these bands behave as the temperature approaches cryogenic levels. This exploration includes a discussion on the formation of the bands, a process linked to the electronic structure of semiconductors. Furthermore, the variation of effective mass with temperature also contributes to this behavior. The electronic structure, energy bands of the semiconductor, nature of the bandgap energy, and dependence of the effective mass on temperature are discussed in this chapter.

Electronic Structure

Silicon crystallizes into a structure conceptualized as "two overlapping face centered cubic lattices," each with a side length measuring 0.543 nm (or 5.43 Å) [114], as depicted in Fig. 3.1. Atoms within silicon have different electron energy levels, known as (permissible) energy bands. As atoms converge, the energy levels begin to intermingle and overlap. Bringing N atoms closer results in the formation of N discrete energy levels. In a large scale crystal lattice, these levels coalesce to

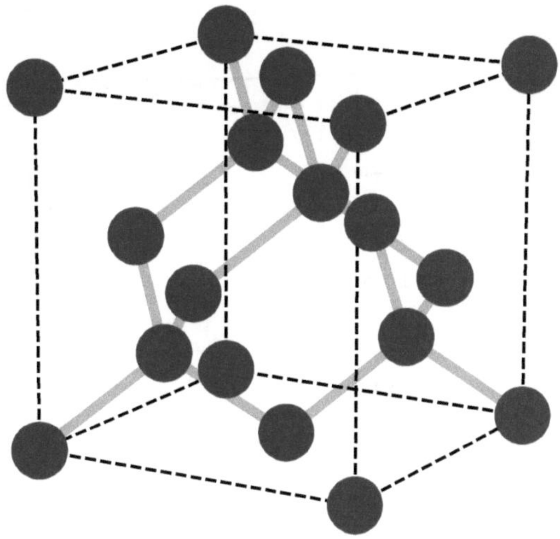

Fig. 3.1 A silicon crystal lattice is illustrated as "two interlocking face-centered cubic lattices," each with a side length of 0.543 nm

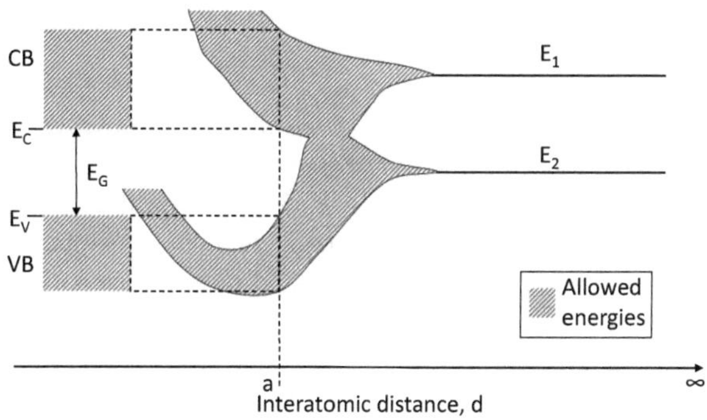

Fig. 3.2 Formation of energy bands within a silicon crystal lattice as a function of the interatomic distance *d*

form continuous energy bands. The permissible energy bands within the crystal lattice as a function of the interatomic distance are illustrated in Fig. 3.2. At greater interatomic distances, the highest identified energy levels are E_1 and E_2. As the atoms become closer, these levels merge to create two distinct yet continuous energy bands, recognized as the conduction band (CB) and the valence band (VB), separated by a bandgap (E_G) where no electron states are permitted.

For a more tangible grasp of energy bands, consider a hydrogen atom, a simple entity comprising a single proton with an orbiting electron. The energy level diagram of a hydrogen atom is illustrated in Fig. 3.3. Here, the electron may occupy distinct energy levels from $n = 1$ to $n = \infty$, with each level demanding a specific energy for the electron to escape the atomic nucleus. This liberation energy is

$$E = -\frac{13.56}{n^2}. \tag{3.1}$$

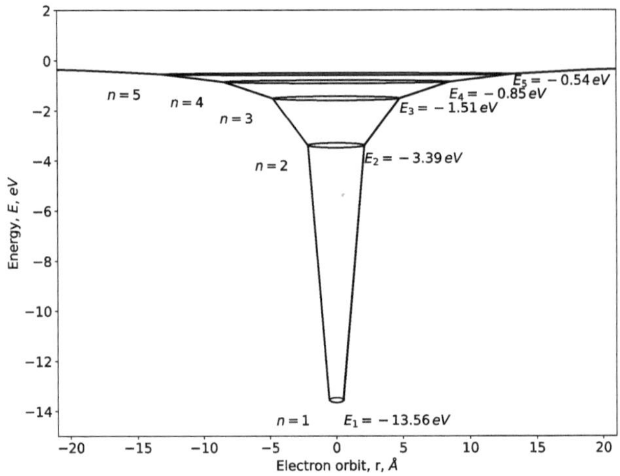

Fig. 3.3 Energy level diagram of a hydrogen atom, comprising a single electron circling a solitary proton

For the first level where $n = 1$, an electron requires an energy of 13.56 eV to become free. Excitation to higher energy states diminishes the energy required for liberation. The initial five energy levels, n = 1 to 5, are illustrated in Fig. 3.3.

Introducing an n-type dopant such as phosphorus into silicon results in the appearance of electrons at elevated energy levels. For instance, a doping concentration of N atoms/cm^3 produces N electrons within a cubic centimeter of silicon residing at these higher energy states, each electron needing only 0.045 eV to transition into the conduction band. This energy can be sourced from phonons, the vibrational energy within the lattice. At standard room temperature, the energy provided by the phonons is sufficient for these electrons to enter the conduction band, rendering the doped silicon conductive. However, as the temperature declines, the phonon energy no longer is sufficient to excite all of the dopant atoms. Below 30 K, virtually all of the electrons are confined within potential wells, and the doped silicon transitions to an insulating state.

The energy distribution in a crystalline structure is characterized by Schroedinger's equation [115]. This equation describes the energy and momentum dynamics of particles transitioning through a crystal lattice. Solutions to Schroedinger's equation yield insight into the free carrier density, mobility, potential, and other particle characteristics, collectively referred to as a *wavefunction*. Electrons, among other particles, are influenced by the periodic potential generated by the lattice atoms, as exemplified in Fig. 3.4.

This periodic potential, denoted as $V_{(\vec{r})}$, aggregates a series of periodic potentials and is modeled as

$$V(\vec{r}) = V\left(\vec{r} + l\vec{a} + m\vec{b} + n\vec{c}\right), \tag{3.2}$$

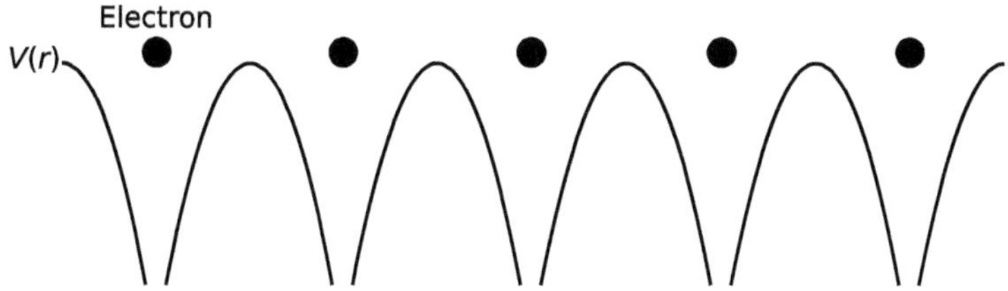

Fig. 3.4 Periodic potential in a one-dimensional lattice

Fig. 3.5 First Brillouin zone of the silicon lattice [116]

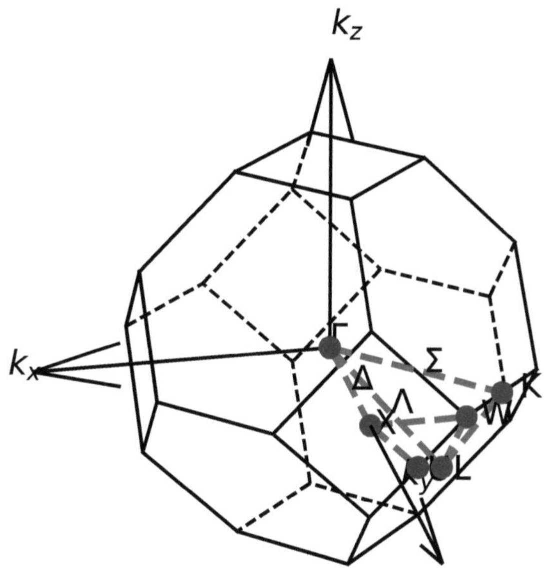

where a, b, and c represent primitive basis vectors, l, m, and n are integers, and r signifies the distance. The Schroedinger's equation is formulated as

$$-\frac{\hbar^2}{2m_0\nabla^2 + V(\vec{r})}\Psi(\vec{r}) = E\Psi(\vec{r}), \tag{3.3}$$

where $\Psi(\vec{r})$ represents a *wavefunction*, E is the energy eigenvalue, $\hbar$ is the reduced Planck constant, and ∇^2 signifies the Laplacian.

Electron momentum p is analogous to a wave with a wavelength $\lambda = \frac{h}{p}$ or a wave vector $\vec{k} = \frac{\vec{p}}{\hbar}$. This wave vector is characterized by the components of electron momentum, $p_x = h_{k_x}$, $p_y = h_{k_y}$, and $p_z = h_{k_z}$. The reciprocal lattice represents a lattice in terms of waves or as the Fourier transform of the crystal lattice. The *first Brillouin zone* of this reciprocal lattice, corresponding to the silicon lattice portrayed in Fig. 3.1, is depicted in Fig. 3.5.

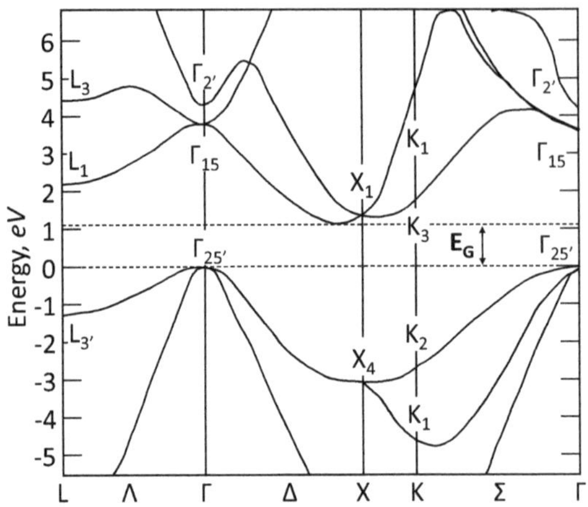

Fig. 3.6 Electronic energy band structure of silicon

Solving Schroedinger's equation provides the Bloch function,

$$\Psi(\vec{r}) = U_n\left(\vec{k}, \vec{r}\right) \cdot e^{j\vec{k}\cdot\vec{r}}, \tag{3.4}$$

where U_n is a periodic function in terms of $\vec{r}$, and n is the band index. The energy band structure of silicon can be described as $\Gamma = (0, 0, 0)$ at the center of the cell. For axes <111>(Λ), <100>(Δ), <110>(Σ), and the intersections of these axes with cell edges, the band structure is, respectively, $L = \frac{2\pi}{a}(1/2, 1/2, 1/2)$, $X = \frac{2\pi}{a}(0, 0, 1)$, and $K = \frac{2\pi}{a}(3/4, 3/4, 0)$ [117]. The energy band diagram for silicon, an indirect transition band material, indicated by the displacement of the upper valence band Γ_{25} from the lower conduction band χ_1, is illustrated in Fig. 3.6 [118, 119]. Nonetheless, if an electron is endowed with approximately 3 eV of energy, a direct transition between the valence band Γ_{25} and Γ_{15} is feasible.

Conduction Band, Valence Band, and Energy Gap

The bandgap energy E_G is depicted in Fig. 3.6 as the region devoid of electronic states. Those energy bands located above this gap are termed the conduction bands, while the bands below the gap are referred to as the valence bands. The energy-wave vector ($E - k$) relationship for regions near the minimum of the conduction band and the maximum of the valence band is captured by the following set of equations [120],

$$E(\vec{k}) = \frac{\hbar^2}{2m}k^2, \tag{3.5}$$

$$E(\vec{k}) = \frac{\hbar^2}{2}\left(\frac{k_l^2}{m_l*} + \frac{k_t^2}{m_t*}\right), \tag{3.6}$$

and

$$E(\vec{k}) = ak^2\left(1 \pm g(\Omega_p, \Omega_a)\right). \tag{3.7}$$

Equation (3.5) illustrates the simplest scenario, known as the *spherical* case, where the electron mass is considered a scalar. This approximation is applied at the Γ point (center of the first Brillouin zone, as noted in Figs. 3.5 and 3.6) for the minima of the conduction band and the maxima of the split-off valence band [121]. For the *ellipsoidal* case, described by (3.6), the electron mass is treated as a tensor with distinct longitudinal and transverse components, respectively, m_1^* and m_2^*. This model is particularly relevant at the L minima (center of the hexagon shown in Fig. 3.6) along the Δ direction (from Γ point to the center of the square face shown in Fig. 3.6). Equation (3.7) pertains to the *warped* case, which can be applied at the two degenerate maxima of the valence band, where ± represents the case for heavy and light holes. The angles Ω_p and Ω_a denote, respectively, the polar and azimuthal angles relative to the crystal axes. The function g is [122]

$$g(\Omega_p, \Omega_a) = \sqrt{b^2 + c^2\left[sin^4\Omega_p cos^2\Omega_a sin^2\Omega_a + sin^2\Omega_p cos^2\Omega_p\right]}, \tag{3.8}$$

with parameters a, b, and c,

$$a = \frac{\hbar|A|}{2m_0}, \tag{3.9}$$

$$b = \frac{|B|}{|A|}, \tag{3.10}$$

$$c = \frac{|C|}{|A|}, \tag{3.11}$$

and m_0 is the rest mass of an electron, specified as 9.1×10^{-31} kg. The inverse mass parameters, $A = -4.27$, $B = -0.63$, and $C = 4.93$, are determined from cyclotron resonance measurements at 4 K [122, 123]. The parabolic representation of the silicon band structure is shown in Fig. 3.7, with the conduction band minima positioned at the Γ point ($\vec{k} =< 0, 0, 0 >$) along the Δ lines ($\vec{k} =< 1, 0, 0 >$) and at the L points ($\vec{k} =< \pi/a_0, \pi/a_0, \pi/a_0 >$). The peak of the valence band is at the Γ point, where two bands converge and a third band splits off due to spin-orbit interactions (see Fig. 3.7).

Application of high electric fields results in carrier energies far from the band edges, displaying nonparabolic behavior. $k \cdot p$ perturbation theory offers a model for this condition based on a two band model [124],

$$E(1 + \alpha E) = \frac{\hbar^2 k^2}{2m^*} = \gamma(k), \tag{3.12}$$

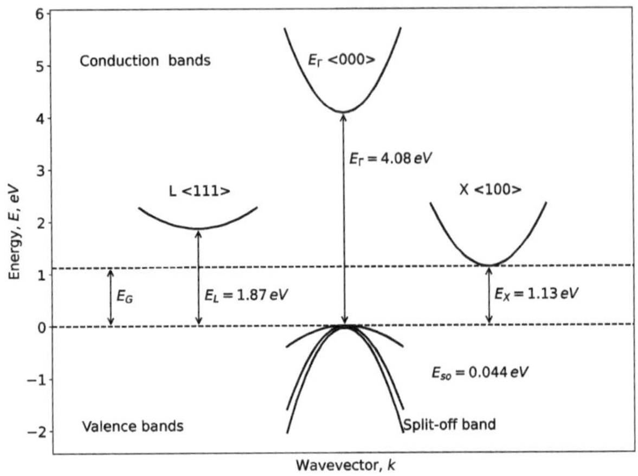

Fig. 3.7 Parabolic approximation of silicon band structure

where α is the nonparabolicity parameter,

$$\alpha = \frac{1}{E_G}\left(1 - m^*/m_0\right)^2 . \tag{3.13}$$

As the temperature of a semiconductor drops, the phonon energies become less, enlarging both the direct and indirect bandgaps. A lower temperature leads to thermal contraction of the lattice, reducing the lattice constant a_0, thereby altering the bandgap. Concurrently, at lower temperatures, lattice vibrations are decreased, impacting neutron and X-ray scattering, and hence smoothing the periodic potential (see Fig. 3.4). This process alters the energy gap—an effect known as the Debye-Waller factor or B factor. Additionally, the lower phonon energy weakens the electron-phonon coupling, reducing the repulsion between electron states within the same band (intraband) and between different bands (interband) due to second-order perturbations—referred to as Fan terms [124, 125].

The impact of a reduced temperature on the energy gap E_G is due to changes in the phonon frequencies [126],

$$E_G = E_{G0} + kT \sum_{q,b} ln \frac{sinh\left[\hbar\omega_u(\vec{q})/(2k_B T)\right]}{sinh\left[\hbar\omega_l(\vec{q})/(2k_B T)\right]}, \tag{3.14}$$

where $\vec{q}$ is the phonon wavevector, and $\omega(\vec{q})$ represents the angular frequency. The sum covers all $\vec{q}$ and phonon branches b [126]. $\omega_u(\vec{q})$ and $\omega_1(\vec{q})$ correspond to the angular frequency where the electron resides, respectively, in the upper and lower energy states. The basic electronic bandgap E_{G0}, unaffected by external influences, is 1.1702 eV [104]. At lower temperatures, an electron excitation causes $\omega_n(\vec{q}) > \omega_1(\vec{q})$, increasing the bandgap energy E_G.

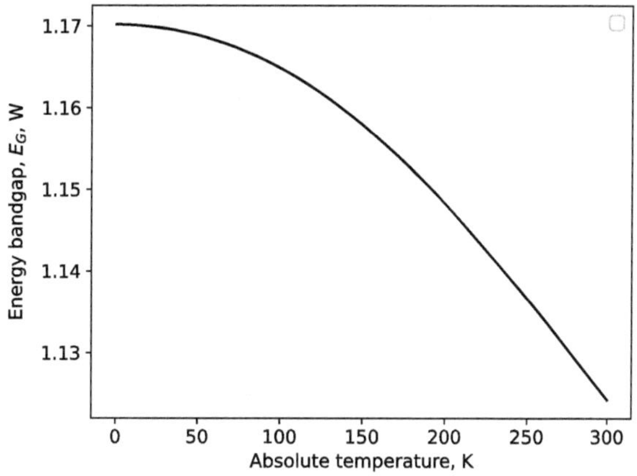

Fig. 3.8 Silicon energy bandgap as a function of temperature [127]

The silicon bandgap has been empirically determined across temperatures ranging from 2 K to 300 K [127], with an approximation to within an error margin of less than 0.05% possible using a fourth order polynomial [106],

$$E_G = E_{G0} + E_{G1}T + E_{G2}T^2 + E_{G3}T^3 + E_{G4}T^4. \tag{3.15}$$

$E_{G0} = 1.1702$, $E_{G1} = -3.6277 \times 10^{-6}$, $E_{G2} = -3.9703 \times 10^{-7}$, $E_{G3} = 1.3207 \times 10^{-9}$, and $E_{G4} = 3.2798 \times 10^{-12}$. The relationship between the energy bandgap of silicon and temperature is illustrated in Fig. 3.8. The bandgap increases with decreasing temperature, as expected from the abovementioned effects.

Effective Mass

The effective mass of a semiconductor is intimately linked to the $E - k$ vector relationship, as previously indicated. From (3.5), the effective mass is

$$\frac{1}{m^*} = \frac{1}{\hbar^2} \cdot \frac{\partial^2 E}{\partial k^2}. \tag{3.16}$$

In the conduction band of silicon, which is ellipsoidal in shape, two distinct masses exist, the longitudinal electron mass and transverse electron mass. These masses at 4.2 K are [123]

$$m_l^* = \frac{m_l}{m_0} = 0.9163 \pm 0.0004, \tag{3.17}$$

$$m_t^* = \frac{m_t}{m_0} = 0.1905 \pm 0.0001. \tag{3.18}$$

The density-of-states effective mass, denoted as m_{de}, is determined by employing the geometric mean at the minimum point of the conduction band. With six minima present within the conduction band of silicon, m_{de} is

$$m_{de}^{*} = (6\sqrt{m_l^{*}(m_t^{*})^2})^{2/3}. \tag{3.19}$$

Contrary to the typical assumption of temperature independence, experimental findings confirm that the mass is indeed temperature dependent [106]. Two distinct categories of effective conduction mass are recognized according to the orientation of a crystal: m_{ce} for electrons and m_{ch} for holes. The effective conduction mass of electrons m_{ce}^{*} is

$$\frac{1}{m_{ce}^{*}} = 1/3\left(\frac{1}{m_l^{*}} + \frac{2}{m_t^{*}}\right), \tag{3.20}$$

where subscripts l and t denote, respectively, the longitudinal and transverse scenarios, stemming from the elliptical nature of the $E - k$ curve and the resulting curvature of the band structure.

For holes, the effective conduction mass m_{ch}^{*} incorporates three different mass types corresponding to the heavy hole band m_{hh}, light hole band m_{lh}, and the split-off band m_{soh}^{*},

$$\frac{1}{m_{ch}^{*}} = 1/3\left(\frac{1}{m_{hh}^{*}} + \frac{1}{m_{lh}^{*}} + \frac{1}{m_{soh}^{*}}\right). \tag{3.21}$$

Additionally, the effective density-of-states mass for electrons is included in (3.16), whereas for holes, m_{dh}^{*} is

$$m_{dh}^{*} = \left[(m_{hh}^{*})^{3/2} + (m_{lh}^{*})^{3/2} + (m_{soh}^{*})^{3/2}\right]^{2/3}. \tag{3.22}$$

The carrier mobility is affected by the effective conduction mass, while the density of states is affected by the effective density-of-states mass. The longitudinal effective mass remains unaffected by the doping concentration [128], but the effective density-of-states mass varies with temperature [104] and increases for doping concentrations exceeding 1×10^{18} cm^3. The electron density-of-states effective mass m_{de}^{*} is graphically presented in Fig. 3.9, exhibiting a 33% increase at 100 K when the doping levels are elevated from 5×10^{17} to 1×10^{20} cm^3 [128–130].

The transverse effective mass of electrons m_t^{*} is depicted in Fig. 3.10, based on experimental data [104] and expressed by the following equation,

$$m_t^{*} = a_{e0} + a_{e1}T + a_{e2}T^2 + a_{e3}T^3 + a_{e4}T^4, \tag{3.23}$$

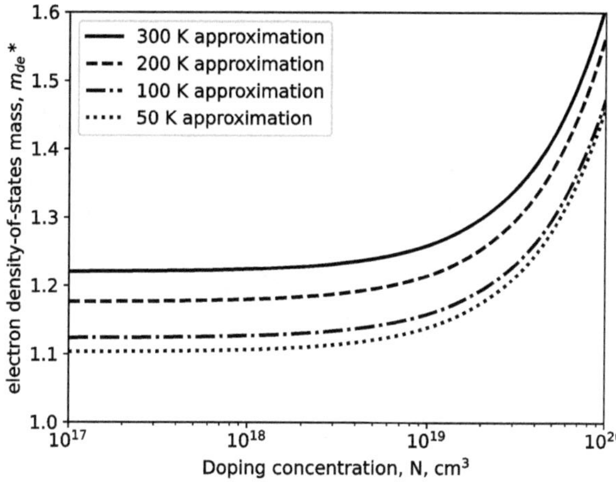

Fig. 3.9 Electron density-of-states mass of silicon at different temperatures [104]

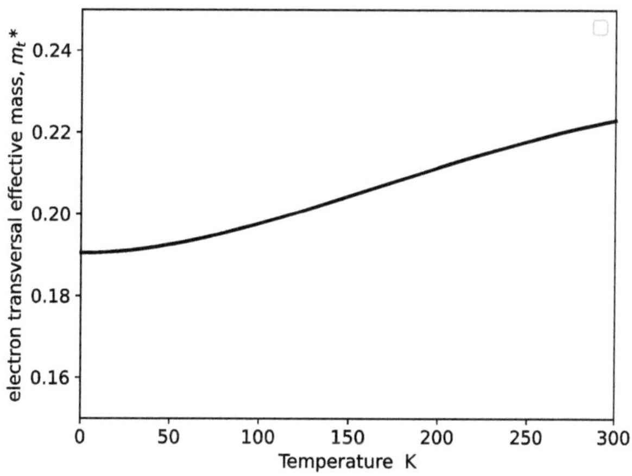

Fig. 3.10 Electron transversal effective mass as a function of temperature

Table 3.1 Measured effective mass values at 4.2 K [104]

m_l^*/m_0	0.9163
m_t^*/m_0	0.1905
m_{hh}^*/m_0	0.5300
m_{lh}^*/m_0	0.1550
m_{so}^*/m_0	0.2350

with constants $a_{e0} = 0.19049$, $a_{e1} = -2.0905 \times 10^{-6}$, $a_{e2} = 9.8985 \times 10^{-7}$, $a_{e3} = -2.6798 \times 10^{-9}$, and $a_{e4} = 2.0270 \times 10^{-12}$ [104]. The effective mass for electrons in silicon at 4.2 K is listed in Table 3.1.

Assuming a constant m_l^* [104] and utilizing the formulations provided in (3.19) and (3.20), the effective density-of-states mass m_{de}^* and the effective conduction mass m_{ce}^* for varying temperatures are displayed in Fig. 3.11.

Assessing the effective mass of holes is more complicated than electrons due to the nonparabolic nature of the valence bands [104]. Alterations in the lattice constant

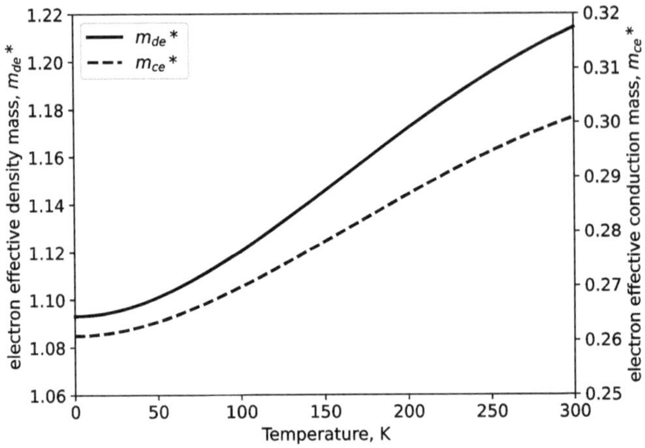

Fig. 3.11 Electron effective density-of-states mass m^*_{de} and effective conduction mass m^*_{ce} as a function of temperature [104]

and electron-phonon interactions at different temperatures add to the complexity of the effective mass within the valence band.

Although prevailing theories have suggested that the heavy hole effective mass for silicon remains unchanged across different temperatures [131], recent experimental evidence indicates that the mass is temperature sensitive [132]. The effective mass of the light hole band is

$$m^*_{lh} = \left[\frac{(m'_{lh})^{3/2}}{F_{1/2}E_F} \int_0^{\xi/(kT)} \frac{\sqrt{E}}{1+e^{E-E_F}} dE + \frac{(m'_{hh})^{3/2}}{F_{1/2}E} \int_{\xi/(kT)}^{\infty} \frac{\sqrt{E-E_1}}{1+e^{E_1-\eta_1}} dE_1 \right]^{2/3}, \tag{3.24}$$

where m'_{lh} is at 4.2 K, $F_{1/2}$ signifies the Fermi-Dirac integral with additional parameters $\xi = 0.02eV$, $\zeta = \xi - E_{so}/3$, E_F is the Fermi level, $E_1 = E - \frac{E_{so}}{3kT}$, and $\eta_1 = E_F - \frac{E_{so}}{3kT}$. The effective mass of the split-off band m^*_{so} is

$$m^*_{soh} = \frac{(m'_{soh})^{3/2}}{F_{1/2}E_F} \int_0^{\infty} \frac{\sqrt{E_2}}{1+e^{E_2-\eta_2}} dE_2, \tag{3.25}$$

with the associated energy terms, $E_2 = E - \frac{E_{so}}{kT}$ and $\eta_2 = E_F - \frac{E_{so}}{kT}$.

Nevertheless, for computational convenience, the effective mass of the heavy hole, light hole, and split-off bands are often approximated using polynomial expressions,

$$m^*_h = 0.51741 + 2.5139 \times 10^{-3} - 4.4117 \times 10^{-6} + 2.6974 \times 10^{-9}, \tag{3.26}$$

$$m^*_l = 0.14615 + 3.7414 \times 10^{-4} - 1.8809 \times 10^{-7}, \tag{3.27}$$

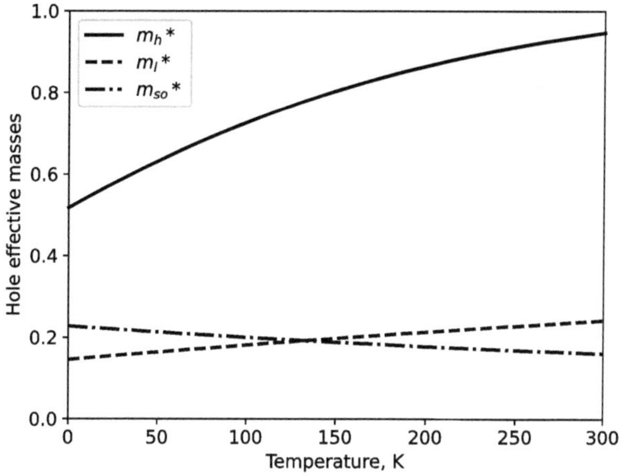

Fig. 3.12 Polynomial approximation of the heavy hole effective mass m_h^*, light hole effective mass m_l^*, and split-off band effective mass m_{so}^* as temperature varies [104]

and,

$$m_{so}^* = 0.22775 - 2.963 \times 10^{-4} + 2.3872 \times 10^{-7}. \tag{3.28}$$

These mass estimates are shown in Fig. 3.12, highlighting the increase in the heavy hole and light hole effective mass with temperature, while noting a reduction in the split-off band effective mass.

3.1.2 *Free Charge Density and Fermi Level*

Fermi Level

The exclusion principle formulated by Pauli posits that within a given quantum state, there cannot exist more than one particle with half-integer spins [133]. Emerging from this principle, Fermi-Dirac statistics describe the distribution of electrons and holes across energy states within a system, extending beyond the single atom application of Pauli's principle [133]. The Fermi-Dirac formulation quantifies the probability that an energy state at equilibrium E is occupied,

$$f(E) = \frac{1}{1 + \exp{(E - E_F)/kT}}, \tag{3.29}$$

where E_F denotes the Fermi level. This equation is pertinent to energy states within a narrow band of a few kT from E_F. For energy significantly above the Fermi level, classical distribution prevails. Despite a natural inclination for charge carriers to first fill the lower energy states, a non-zero probability exists for occupancy of higher states even at absolute zero temperature. The Fermi level acts as a benchmark, normalizing the energy states within a material, with those states beneath the benchmark (Fermi level) more likely occupied and those above the Fermi level more likely vacant. The Fermi level is determined by the concentration of charge carriers per unit volume n_0, and the density-of-states function at specific energy levels $g(E)$. The aggregate of charge carriers per unit volume is [104]

$$n_0 = \int_{-\infty}^{\infty} f(E) \cdot g(E)\, dE. \tag{3.30}$$

The introduction of impurities into a semiconductor introduces new energy levels, such as E_d, associated with donor atoms. This doping results in two states per spin, termed a degenerate spin [134]. Nevertheless, once one of these quantum states is filled, the second state becomes unavailable, reflecting that the ionized impurity atom is satisfied by a single electron. Consequently, the presence of impurities modifies both the distribution function and the anticipated behavior of the charge carriers.

Electron and Hole Population

With a decrease in temperature, fewer electrons populate the conduction band, and more states in the valence band become filled. Conversely, as the temperature increases, electrons fill the conduction band, creating holes in the valence band, both of which are vital for electrical conductivity. The band structure of covalent semiconductors like silicon or germanium, illustrated in Fig. 3.13, demonstrates that as atoms approach each other, the energy bands first overlap at point A and diverge at point B. To the left of point A, the valence bands are composed of both *s* and

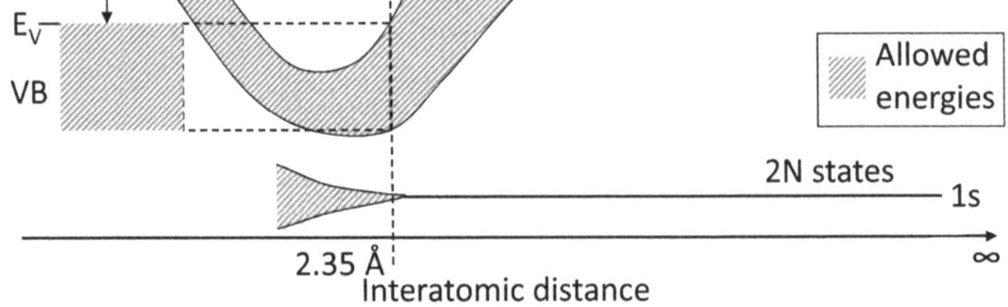

Fig. 3.13 Energy band of silicon as a function of the lattice constant a_0

p wavefunctions. Beyond point B, the carriers must be excited into the conduction band for a semiconductor to conduct. This excitation of electrons can be induced through thermal effects, optical radiation, or by applying an electric field, leading to the generation of electron and hole populations, which constitute the free charge carriers in the material.

Charge Carriers in Intrinsic Silicon

In intrinsic silicon, which is silicon in pure form without any doping elements, the number of electrons equals the number of holes. The free carrier population in such a material is governed by the Fermi-Dirac distribution [135, 136]. Consequently, the distribution function for the energy states is akin to that of a free particle. For the conduction band, where $E > E_c$, the distribution function is represented as

$$g_c(E)dE = \left(\frac{8\sqrt{2}\pi}{h^3}\right)(m_n*)^{3/2}\sqrt{E - E_c}dE. \tag{3.31}$$

For the valence band, where $E < E_v$, the distribution function is similarly described [104],

$$g_v(E)dE = \left(\frac{8\sqrt{2}\pi}{h^3}\right)(m_p*)^{3/2}\sqrt{E_v - E}dE. \tag{3.32}$$

The quantity of free electrons per unit volume in the conduction band, denoted as n_0 for energy level E within a differential energy range dE, is [104]

$$d_{n_0} = f_0(E)g_c(E)dE. \tag{3.33}$$

f_0 is the Fermi function, and g_c is the density of states. At room temperature, $kT = 0.0259$ eV. Since the silicon bandgap is about 1.12 eV and $E_G \gg kT$, the Fermi distribution function can be simplified to

$$f_0(E) \approx e^{-(E-E_F)/(kT)}, \tag{3.34}$$

which is known as the Maxwell-Boltzmann approximation [137]. By integrating (3.33), the total number of electrons per volume in a conduction band is

$$n_0 = \left(\frac{8\sqrt{2}\pi}{h^3}\right)(m_n*)^{3/2}e^{E_F/(kT)}\int_{E_c}^{\infty}\sqrt{E_c - E}e^{-E/(kT)}\,dE. \tag{3.35}$$

By substituting $y_c = E - \frac{E_c}{kT}$,

$$n_0 = \left(\frac{8\sqrt{2\pi}}{h^3}\right)(m_n * kT)^{3/2} e^{E_F - E_c/(kT)} \int_0^{\infty} \sqrt{y_c} e^{-y_c} \, dy_c. \tag{3.36}$$

By substituting the integral with gamma function $\Gamma(3/2) = \sqrt{\pi}/2$, (3.36) reduces to

$$n_0 = U_c e^{-(E_c - E_F)/(kT)}, \tag{3.37}$$

where U_c is the effective density of states in the conduction band,

$$U_c = 2\left(2\pi m_n * (kT/h^2)\right)^{3/2}. \tag{3.38}$$

Similarly, the total number of holes in the valence band per unit volume p_0 is

$$p_0 = U_v e^{-(E_F - E_v)/(kT)}, \tag{3.39}$$

where U_v is the effective density of states in the valence band,

$$U_v = 2\left(2\pi m_p * (kT/h^2)\right)^{3/2}. \tag{3.40}$$

The intrinsic concentration n_i is [104]

$$n_i = \sqrt{n_0 \cdot p_0} = 2\left(\frac{2\pi\sqrt{m_n * m_p *} kT}{h^2}\right)^{3/2} \cdot e^{-E_G/(2kT)}. \tag{3.41}$$

The dependence of the intrinsic concentration n_i of silicon on temperature is shown in Fig. 3.14. For intrinsic semiconductors, the position of the Fermi level E_F is

$$E_F = 1/2(E_c + E_v) + kt \times ln(m_p * / m_n *)^{3/4}. \tag{3.42}$$

Charge Carriers in Doped Silicon

In doped semiconductors, each energy level is paired with two quantum states, termed a degenerate state [134]. Consequently, the Fermi energy is not the sole determinant of the statistical behavior in doped silicon. As previously mentioned, the occupancy of one degenerate state precludes the availability of the second state, as the electron satisfies the valence requirement. Additionally, doped semiconductors feature energy levels specific to donors or acceptors. By convention, a donor level is considered neutral when the donor energy level includes an electron,

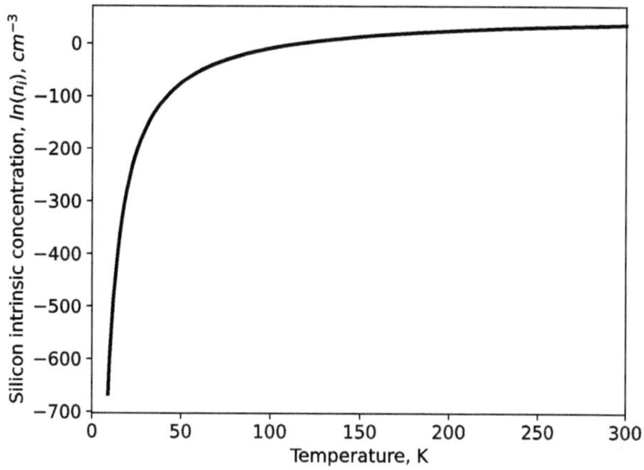

Fig. 3.14 Intrinsic concentration of silicon n_i as a function of temperature, as described by (3.41)

whereas an acceptor level is neutral when the acceptor energy level is devoid of electrons. Conversely, a donor level is negatively charged when vacated by an electron, and an acceptor level is negatively charged when filled with an electron. The Fermi level must maintain charge neutrality, meaning the aggregate of negative charges (electrons and ionized acceptors) must be equivalent to the total positive charges (holes and ionized donors),

$$n = N_D^+ + p, \text{ for n-type doped semiconductor,} \tag{3.43}$$

and

$$p = N_A^- + n, \text{ for p-type doped semiconductor.} \tag{3.44}$$

n and p are the electron and hole concentration in, respectively, the conduction and valence bands. N_D and N_A are, respectively, the number of ionized donors and acceptors,

$$N_D^+ = N_D \frac{1}{1 + 2e^{-(E_d - E_F)/(kT)}}, \tag{3.45}$$

and

$$N_A^+ = N_A \frac{1}{1 + 2e^{(E_a - E_F)/(kT)}}. \tag{3.46}$$

N_D and N_A are, respectively, the donor and acceptor concentration added to a crystal, and E_d and E_a are, respectively, the donor and acceptor energy levels. The assumptions for room temperature are

$$N_D \ll 1/2 U_c e^{-(E_c - E_d)/(kT)}, \text{ for donor impurities,} \tag{3.47}$$

and

$$N_A \ll 1/2U_b e^{-(E_a-E_v)/(kT)}, \text{ for acceptor impurities.} \tag{3.48}$$

At cryogenic temperatures, the phenomenon of carrier freezeout invalidates those assumptions that apply only to room temperature. For n-type doped semiconductors, at temperatures below 30 K, the charge neutrality condition is governed by specific parameters that differ significantly from room temperature. The charge neutrality condition for an n-type doped semiconductor dictates that [104]

$$n_0 = N_D - n_d, \tag{3.49}$$

where n_d is the number of non-ionized donors per unit volume,

$$n_d = \frac{N_D}{1 + 1/2e^{(E_d-E_F)/(kT)}}. \tag{3.50}$$

By substituting (3.37) and (3.50) into (3.49),

$$U_c e^{-(E_c-E_F)/(kT)} + 2U_c e^{-(E_c+E_d-2E_F)/(kT)} - N_D = 0. \tag{3.51}$$

By using substitutions $\alpha_c = e^{-E_c/(kT)}$, $\alpha_d = e^{-E_d/(kT)}$, and $\alpha_F = e^{E_F/(kT)}$, (3.51) can be rewritten as a quadratic expression,

$$\alpha_F^2 + \frac{1}{2\alpha_d} \cdot \alpha_F - \frac{N_d}{2U_c\alpha_c\alpha_d} = 0. \tag{3.52}$$

Solving (3.52) for E_F,

$$E_F = kTln\left(-\frac{1}{4\alpha_d} + \sqrt{\frac{1}{16\alpha_d^2} + \frac{N_D}{2U_c\alpha_c\alpha_d}}\right), \tag{3.53}$$

or

$$E_F = 1/2(E_c + E_d) + kT/2ln\left(\frac{N_D}{2U_c}\right) - kTsinh^{-}1\left(\sqrt{\frac{U_c}{8N_D}}e^{-(E_c-E_d)/(2kT)}\right). \tag{3.54}$$

Modifying the doping concentration in a semiconductor can transform certain properties, shifting from insulator-like to metal-like in terms of the electrical conductivity. This transition results from a shift in the energy distribution within the crystal structure. The process of doping a semiconductor to induce metal-like conductivity is referred to as degeneration. Degeneration takes place when the Fermi level E_F surpasses the energy level of the conduction band E_c. The density of states in this case is depicted in Fig. 3.15.

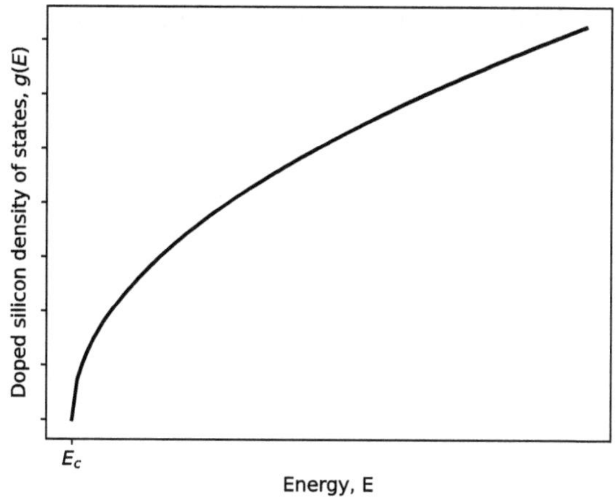

Fig. 3.15 Density of states in a doped silicon as a function of energy E

3.1.3 Carrier Mobility and Carrier Scattering Mechanisms

Applying an electric field to a semiconductor induces carrier drift within the crystal lattice due to several scattering mechanisms. This drift is quantified as carrier mobility, which encompasses several scattering mechanisms such as lattice scattering (phonons), ionized impurity scattering, surface scattering, carrier-to-carrier scattering, neutral impurity scattering, and velocity saturation. Furthermore, carrier mobility is also influenced by the dimensions of the semiconductor. Notably, when the size of a semiconductor is reduced to approximately 100 nm or less, quantum effects start to become significant and must be considered when characterizing the mobility.

Lattice Scattering

Lattice scattering significantly affects mobility among several scattering mechanisms in semiconductors. It occurs due to the interaction of charge carriers, such as electrons or holes, with the vibrational energy of the lattice, known as phonons. The movement of an electron within an applied electric field can be described by the following relation [104],

$$\hbar\vec{k} = m_e^* \cdot \frac{d\vec{v}}{dt} = \vec{F}_0 - m_e^* \frac{\vec{v}}{\tau}, \tag{3.55}$$

where $\vec{k}$ is the wavevector of the phonon, m_e^* is the effective mass of the electron, $\vec{v}$ is the electron velocity, $\vec{F}_0$ is the force exerted by the electric field on the electron, and τ is the relaxation time. The relaxation time is the mean free time of the carrier, which is the average time between each collision. The scattering process effectively acts as a resistance to the movement of electrons, represented by the term $m^* \frac{\vec{v}}{\tau}$. The current density, which is the current per unit area induced by the electric field, is

$$\vec{j} = \vec{I}/A = \frac{\vec{q}_t}{\tau \cdot A} = -ne\vec{v}, \tag{3.56}$$

where q_t represents the total charge, e is the electron charge, and A is the cross-sectional area. Rearranging (3.55) as

$$-m_e^* \frac{\vec{v}}{\tau} = \hbar\vec{k} - \vec{F}_0 = \vec{F}, \tag{3.57}$$

$$-\vec{v} = \tau \cdot \frac{\vec{F}}{m_e^*}, \tag{3.58}$$

and substituting $\vec{v}$ into (3.56), the current density is

$$\vec{j} = -ne\vec{v} = ne \cdot \tau \frac{\vec{F}}{m_e^*} = n \cdot \frac{e\tau}{m_e^*} \cdot e\vec{E}. \tag{3.59}$$

The mobility μ is

$$\mu = \frac{e\tau}{m_e^*}, \tag{3.60}$$

and (3.59) becomes

$$\vec{j} = n \cdot \frac{e\tau}{m_e^*} \cdot e\vec{E} = en\mu\vec{E}. \tag{3.61}$$

However, the mobility requires the phonon energy $\hbar\vec{k}$ and internal energy of the crystal to be determined. Mathematical models have been proposed [138] to approximate the mobility of electrons μ_{pse},

$$\mu_{pse} = \mu_{0ea}(T/300)^{-\alpha_e} + \mu_{0eb}(T/300)^{-\beta_e}, \tag{3.62}$$

and holes μ_{psh},

$$\mu_{psh} = \mu_{0ha}(T/300)^{-\alpha_h} + \mu_{0hb}(T/300)^{-\beta_h}. \tag{3.63}$$

$\mu_{0ea} = 4195cm^2/Vs$, $\mu_{0eb} = 2153cm^2/Vs$, $\mu_{0ha} = 2502cm^2/Vs$, $\mu_{0hb} = 591cm^2/Vs$, $\alpha_e = 1.5$, $\beta_e = 3.13$, $\alpha_h = 1.5$, and $\beta_h = 3.25$. The mobility of electrons and holes as a function of temperature is shown in Fig. 3.16. At lower temperatures, the mobility typically increases due to reduced phonon scattering, which otherwise impedes the drift of the carriers.

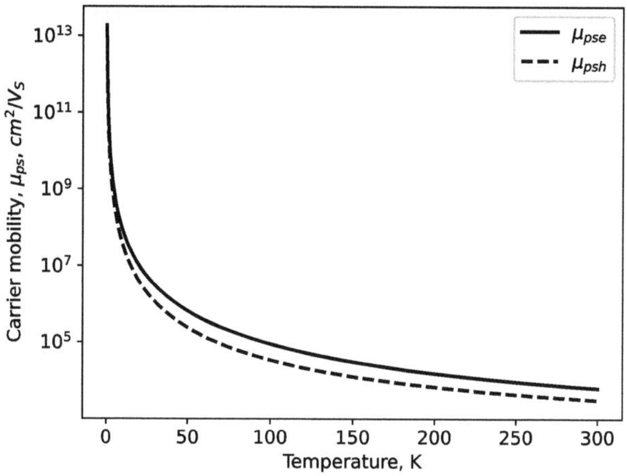

Fig. 3.16 Electron mobility μ_{pse} and hole mobility μ_{psh} due to phonon scattering as a function of temperature [138]

Ionized Impurity Scattering

Ionized impurity scattering in a doped semiconductor affects the carrier mobility by increasing the number of scattering centers, which are the newly introduced atoms [139]. A model describing the mobility influenced by ionized impurity scattering for impurity concentrations below 10^{19} is [139, 140]

$$\mu_{iie} = \sqrt{\frac{\pi}{m_e^*}} \cdot \frac{64\varepsilon^2(2kT)^{3/2}}{e^3 N_{ii}} \cdot \frac{1}{ln(1+x) - x/(1+x)}, \tag{3.64}$$

where the number of ionized impurity atoms is denoted by N_{ii}, and x is

$$x = \frac{24m_e^*\varepsilon(kT)^2}{e^2\hbar^2(n+p)}. \tag{3.65}$$

For hole mobility, the electron effective mass m_e^* should be replaced with the hole effective mass m_h^*. Although these models effectively predict changes in mobility due to impurity scattering, the models cannot be directly combined with the phonon scattering mobility approximations from (3.62) and (3.63). Since phonon scattering and ionized impurity scattering are interdependent, a simple summation is not viable. Rather, the combined effects of phonon and ionized impurity scattering on mobility is modeled through [141]

$$\mu_{piisc} = \mu_{minc} + \frac{\mu_{psc} - \mu_{minc}}{1 + \left(\frac{N_{ii}}{N_{refc}}\right)^{\alpha_{ac}}}, \tag{3.66}$$

where μ_{piisc} is the carrier mobility, subscript c in all of the terms refers to the carrier type, and the carrier type can be either an electron e or hole h. μ_{mine} and μ_{minh} are, respectively,

$$\mu_{mine} = 197.17 - 45.505 logT, \tag{3.67}$$

$$\mu_{minh} = 110.90 - 25.597 logT. \tag{3.68}$$

N_{refe} and N_{refh} are, respectively,

$$N_{refe} = 1.12 \times 10^{17} \cdot (T/300)^{3.2}, \tag{3.69}$$

$$N_{refh} = 2.23 \times 10^{17} \cdot (T/300)^{3.2}, \tag{3.70}$$

and α_{ac} is

$$\alpha_{ac} = \alpha_{ae} = \alpha_{ah} = 0.72 \cdot (T/300)^{0.065}. \tag{3.71}$$

Combining (3.66) with (3.62) and (3.63), the change in mobility due to phonon and ionized impurity scattering can be expressed. The electron mobility and hole mobility, respectively, μ_{piise} and μ_{piish}, due to phonon and ionized impurity scattering, are illustrated in Fig. 3.17, along with the electron mobility and hole mobility, respectively, μ_{pse} and μ_{psh}, due solely to phonon scattering. The doping concentration is $N_{ii} = 1 \times 10^{16}$ cm^{-3}. A decrease in mobility upon the introduction of impurity scattering is revealed in Fig. 3.17. The dependence of the electron

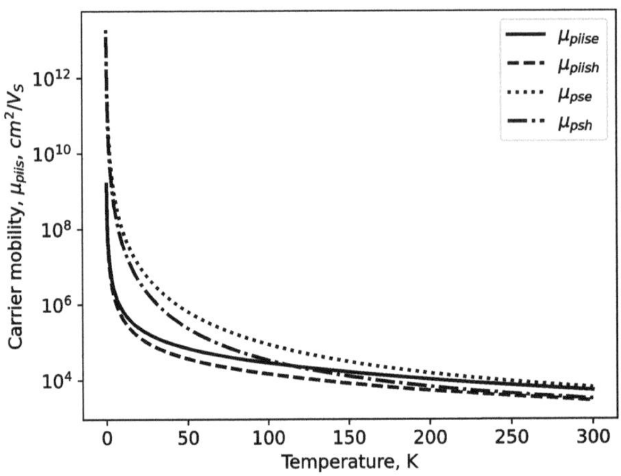

Fig. 3.17 Carrier mobility affected by both phonon and ionized impurity scattering, and carrier mobility influenced by phonon scattering alone as a function of temperature for an impurity concentration of $N_{ii} = 1 \times 10^{16}$ cm^{-3}

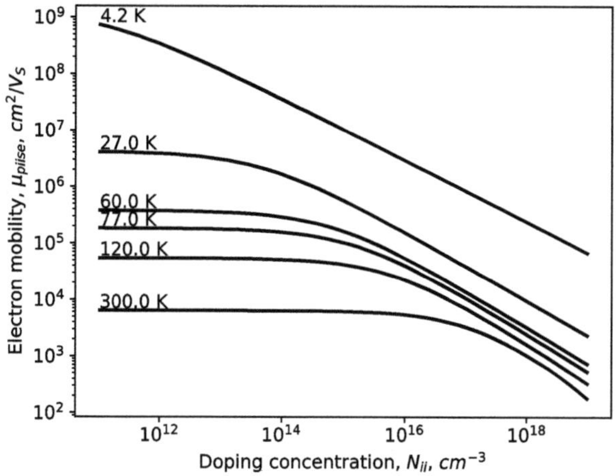

Fig. 3.18 Electron mobility affected by phonon and ionized impurity scattering as a function of doping concentration N_{ii}. The electron mobility is illustrated for six different temperatures, 4.2 K, 27 K, 60 K, 77 K, 120 K, and 300 K

mobility on both phonon and ionized impurity scattering μ_{piise} on the doping concentration N_{ii} is illustrated in Fig. 3.18. The mobility is evaluated at different temperatures: 4.2 K (liquid helium), 27 K (liquid neon), 60 K and 77 K (liquid nitrogen), 120 K (liquid krypton), and 300 K (room temperature).

Velocity Saturation

Electron mobility, incorporating the combined effects of phonon scattering, ionized impurity scattering, and velocity saturation, is described by the following relationship [104],

$$\mu_{piisv,c} = \left(\frac{\mu_{piis,c}}{1 + \left(\frac{\mu_{piis,c} E}{v_{sat,c}} \right)^{\beta_{vs,c}}} \right)^{1/\beta_{vs,c}}, \tag{3.72}$$

where $v_{sat,c}$ denotes the saturation velocity of the carrier. The saturation velocity for electrons $v_{sat,e}$ is

$$v_{sat,e} = 10^7 \cdot (T/300)^{-0.87}, \tag{3.73}$$

and for holes $v_{sat,h}$

$$v_{sat,h} = 8.37 \cdot 10^6 \cdot (T/300)^{-0.52}. \tag{3.74}$$

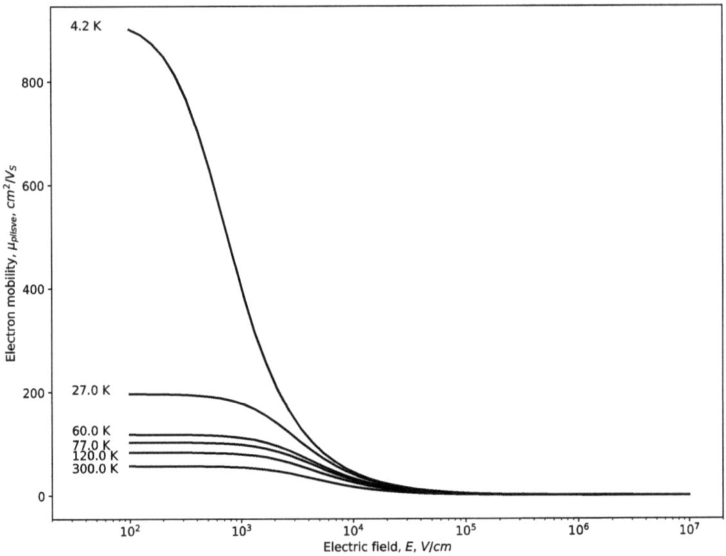

Fig. 3.19 Electron mobility μ_{piisve} as a function of the applied electric field considering phonon scattering, ionized impurity scattering, and velocity saturation. The doping concentration is $N_{ii} = 1 \times 10^{17}$ cm^{-3}

Parameter $\beta_{vs,c}$ is assigned a value of $\beta_{vs,e} = 2$ for electrons and $\beta_{vs,h} = 1$ for holes. A graphical representation of the mobility influenced by phonon scattering, ionized impurity scattering, and velocity saturation is shown in Fig. 3.19. An impurity doping concentration of $N_{ii} = 1 \times 10^{17}$ cm^{-3} is assumed.

Carrier-to-Carrier Scattering

Carrier-to-carrier scattering is an interaction mechanism that occurs when carriers influence the movement of each other. This scattering becomes significant when the carrier concentration surpasses the doping concentration or when the current density is exceptionally high. The variation in mobility due to carrier-to-carrier scattering is [142]

$$\mu_{cc} = \frac{2 \times 10^{17} T^{3/2} / \sqrt{np}}{ln\left(1 + 8.28 \times 10^{8} T^{2} (pn)^{-1/3}\right)}. \tag{3.75}$$

The temperature dependent mobility changes due to carrier-to-carrier interactions, as illustrated in Fig. 3.20. The electron and hole concentration is, respectively, chosen as $n = 1 \times 10^{17}$ cm^{-3} and $p = 1 \times 10^{15}$ cm^{-3}.

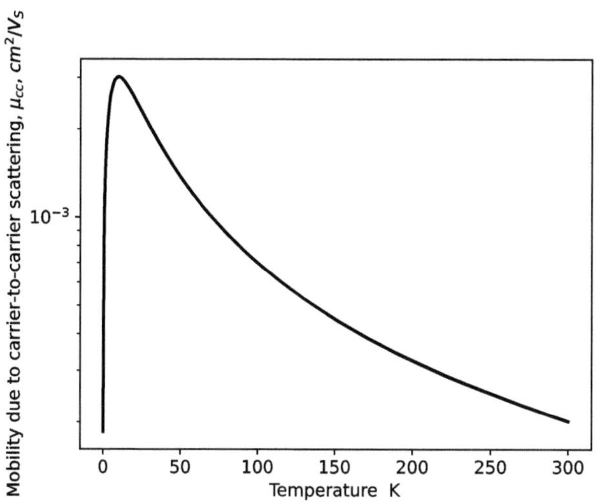

Fig. 3.20 Electron mobility due to carrier-to-carrier scattering μ_{cce} as a function of temperature. The electron and hole concentration is, respectively, $n = 1 \times 10^{17}$ cm^{-3} and $p = 1 \times 10^{15}$ cm^{-3}

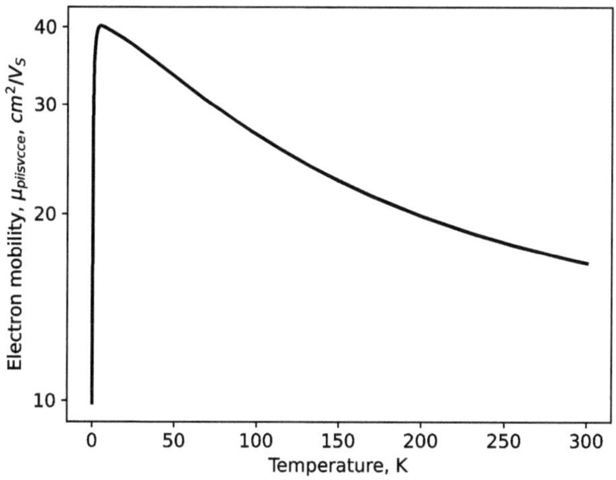

Fig. 3.21 Electron mobility influenced by phonon scattering, ionized impurity scattering, velocity saturation, and carrier-to-carrier scattering $\mu_{piisvcce}$ as a function of temperature

Carrier-to-carrier scattering can be integrated with phonon scattering, ionized impurity scattering, and velocity saturation [104],

$$\mu_{piisvcce} = \frac{1}{1/\mu_{piisve} + 1/\mu_{ce}}. \tag{3.76}$$

The combined effect of these scattering mechanisms on mobility as a function of temperature is shown in Fig. 3.21. The applied electric field is 1×10^4 V/cm. The observed trend indicates a decrease in mobility with increasing temperature, which is attributed to the additional number of carriers at elevated temperatures [104].

Neutral Impurity Scattering

Neutral impurity scattering considers interactions between charge carriers and neutral atoms in doped semiconductors, as opposed to ionized impurity scattering which arises from charged impurity atoms. Neutral impurity scattering is temperature dependent [143, 144] and more prominent at lower temperatures, particularly below 125 K [104]. An early model for electron scattering by neutral impurities was proposed in 1950 as [145]

$$\mu_{niso} = \frac{2\pi^3 e^3 m_e *}{5 N_N \epsilon h^3} \times 10^{-2}, \tag{3.77}$$

where N_N is the density of neutral impurities, the permittivity is $\epsilon_i = 1.03 \times 10^{-12}$ As/Vcm, and h is Planck's constant. This model does not adequately consider temperature dependence, leading to subsequent refinements [143, 146],

$$\mu_{nis} = 0.83 \cdot \mu_{niso} \left[2/3 \sqrt{\frac{kT}{E_N}} + 1/3 \sqrt{E_N} kT \right], \tag{3.78}$$

where μ_{niso} is (3.77), and $E_N = 1.136 \times 10^{-19} \cdot m_e * / m_0 \cdot (\epsilon_0/\epsilon_i)^2$ characterizes the energy related to neutral impurity scattering.

As temperature decreases, ionized impurities increasingly convert to neutral impurities, particularly below 50 K, causing a sharp rise in the neutral impurity concentration and corresponding exponential reduction in the mobility. This effect is revealed in Fig. 3.22. The cumulative electron mobility, which includes the effects of phonon scattering, ionized impurity scattering, velocity saturation, carrier-to-carrier scattering, and neutral impurity scattering, is determined from Matthiessen's rule [147] as

$$\mu_{piisvccnie} = \frac{1}{1/\mu_{piisvcce} + 1/\mu_{nis}}. \tag{3.79}$$

The combined influence of these scattering mechanisms on the electron mobility $\mu_{piisvccnie}$ across a range of temperatures is illustrated in Fig. 3.23.

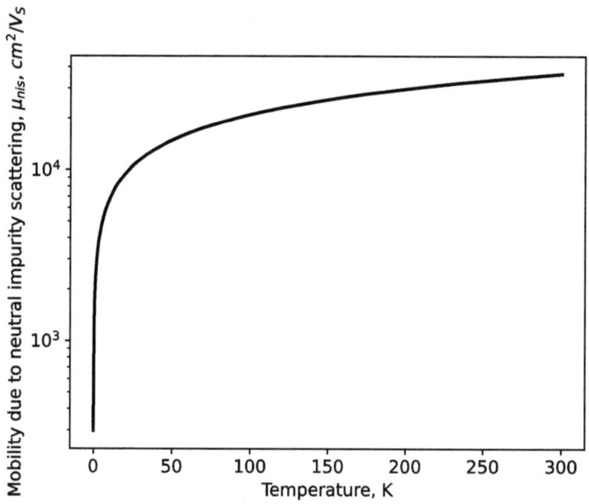

Fig. 3.22 Electron mobility due to neutral impurity scattering μ_{nis} as a function of temperature

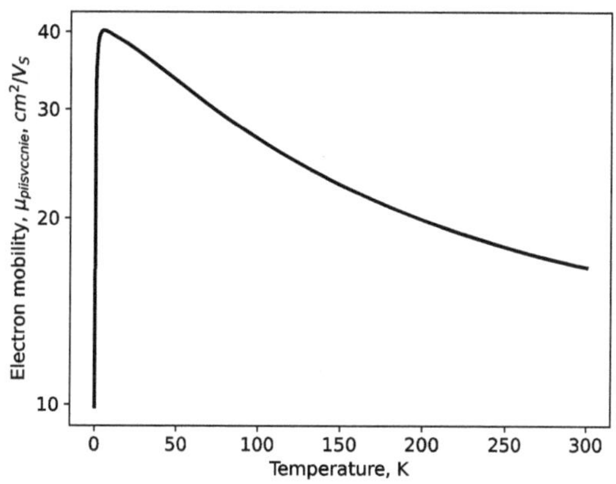

Fig. 3.23 Combined electron mobility incorporating the effects of phonon scattering, ionized impurity scattering, velocity saturation, carrier-to-carrier scattering, and neutral impurity scattering as a function of temperature. The electron and hole concentration is, respectively, $n = 1 \times 10^{17}$ cm^{-3} and $p = 1 \times 10^{15}$ cm^{-3}

Surface Scattering

Surface scattering in semiconductors arises due to irregularities at the material surface, such as roughness, defects, foreign atoms, and dangling bonds [148]. This scattering mechanism affects the mobility of charge carriers near the surface.

A comprehensive mobility model that encompasses the influence of phonon scattering, ionized impurity scattering, velocity saturation, carrier-to-carrier scattering, neutral impurity scattering, and surface scattering is [104, 149]

$$\mu = \mu_{ref,e/h} = \frac{\mu_{ref,e/h} + (\mu_{piisvccnie} - \mu_{ref,e/h}) \cdot [1 - f_r(y)]}{1 + f_r(y) \cdot \left(\frac{S_{e/h}}{S_{ref,e/h}}\right)^{e_{ref,e/f}}}, \tag{3.80}$$

where $\mu_{piisvccnie}$ denotes the mobility resulting from the combined effects of the six phonon scattering effects. The terms $\mu_{ref,e}$ and $\mu_{ref,h}$ are the reference mobilities for electrons and holes, respectively, with the temperature dependence and fitting parameters described as [104]

$$\mu_{ref,e} = 638/a_r \cdot \left(\frac{T}{300}\right)^{-1.19}, \tag{3.81}$$

$$\mu_{ref,h} = 1/a_r \cdot \left(\frac{T}{300}\right)^{-1.09}, \tag{3.82}$$

where a_r, a_t, and a_x are fitting parameters with a value near 1. The function $f_r(y)$ provides a smooth transition from the surface scattering regime to the bulk mobility,

$$f_r(y) = \frac{2e^{-2y/10nm}}{1 + e^{-4y/10nm}}, \tag{3.83}$$

and is depicted in Fig. 3.24. The other parameters are [104]

$$S_{ref,e} = 7 \times 10^5/a_t, \tag{3.84}$$

$$S_{ref,h} = 2.7 \times 10^5/a_t, \tag{3.85}$$

$$e_{ref,e} = 1.3/a_x, \tag{3.86}$$

$$e_{ref,h} = 1.0/a_x. \tag{3.87}$$

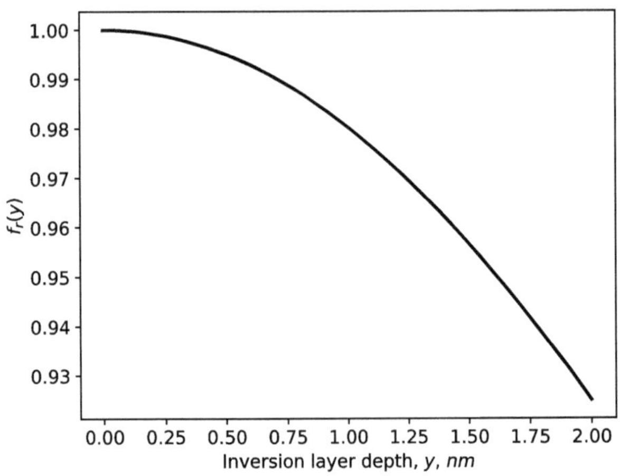

Fig. 3.24 $f_r(y)$ is an arbitrary function describing the transition from surface related effects to bulk properties

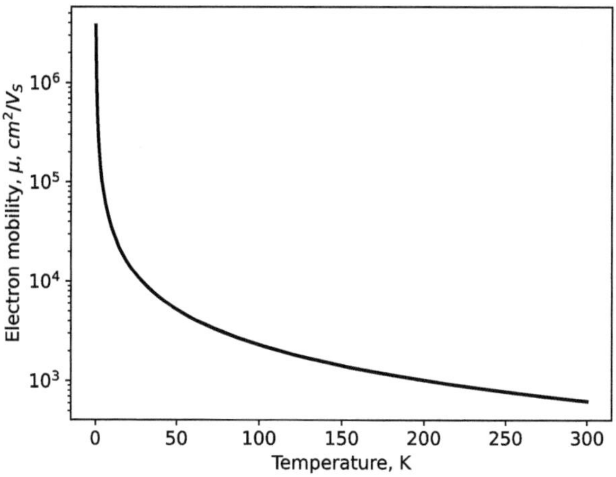

Fig. 3.25 Electron mobility μ, combining the effects of phonon scattering, ionized impurity scattering, velocity saturation, carrier-to-carrier scattering, neutral impurity scattering, and surface scattering, as a function of temperature

$S_{e/h}$ is the maximum electric field perpendicular to the electric flow, known as the interface pressing force [149]. The temperature dependence of the aggregate mobility μ, incorporating all six scattering mechanisms, is shown in Fig. 3.25. The mobility significantly increases at lower temperatures due to the diminished influence of these scattering effects.

3.1.4 Electrical Conductivity

The electrical conductivity of a semiconductor is fundamentally governed by the density of the free carriers n and mobility μ. The conductivity σ is [104]

$$\sigma = e\mu n \frac{W \cdot X_j}{L}, \tag{3.88}$$

where W, X_j, and L are, respectively, the width, junction depth, and length of the conducting channel. The temperature dependence of the electrical conductivity of a semiconductor is influenced by variations in the mobility and carrier density with temperature. Additionally, the doping concentration plays a pivotal role in these parameters.

At low temperatures, a semiconductor may exhibit an insulating behavior due to carrier freezeout, particularly when the doping levels are below the threshold for degeneration [105]. This freezeout effect is typically observed in the steady-state condition (saturated carriers), where the conduction channel is fully established. However, during transient states, as when an electric field is applied, the electrical

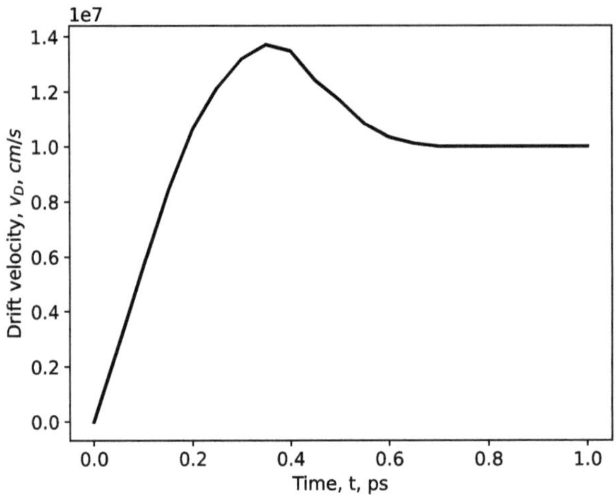

Fig. 3.26 Drift velocity of electrons with an applied electric field of 20 kV/cm [104]

conductivity can deviate from this steady-state behavior due to ballistic transport or velocity overshoot [112]. This phenomenon occurs when the mean free time of the carriers (relaxation time) exceeds the transit time through the channel, resulting in a carrier velocity that surpasses the steady-state drift velocity. The drift velocity of the charge carriers under an applied electric field of 20 kV/cm is shown in Fig. 3.26, where velocity overshoot can be observed [104].

An expression governing the drift velocity of carriers in the ballistic regime is [104]

$$v_D = \frac{eEt}{m^*}, \tag{3.89}$$

where E is the electric field strength and t is the time.

Velocity overshoot is more significant for 'cold' carriers and diminishes as the carriers 'heat up.' The occurrence of ballistic transport requires high frequency electric fields, with the frequency related to the relaxation time τ by $f = 1/\tau$. Simulations suggest a relaxation time ranging from 0.1 to 0.4 picoseconds, corresponding to an electric field frequency in the range of 2.5 to 10 terahertz [104]. The conduction paths suitable for ballistic transport are on the order of the modified de Broglie wavelength [150] $\lambda = \lambda_0 \cdot tanh\,(\varepsilon_r/2kT)$, where λ_0 is between 54 to 108 Å and $\varepsilon_r = 0.063\ eV$, which changes from 4.53 to 10.8 nanometers in the temperature range from 300 K to 4.2 K. At cryogenic temperatures, this behavior indicates that devices with a channel length under 10 nanometers are likely to exhibit ballistic transport at terahertz frequencies.

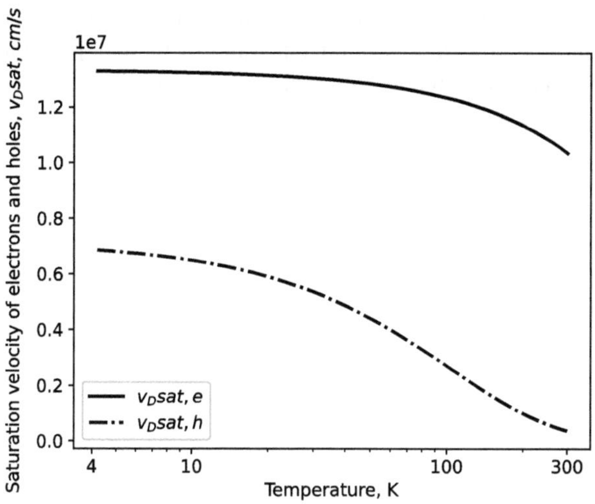

Fig. 3.27 Saturated drift velocity of electrons and holes at different temperatures

The saturated electron velocity $v_{Dsat,e}$ and hole velocity $v_{Dsat,h}$ are modeled, respectively, as [151]

$$v_D sat, e = \frac{2.4 \times 10^7}{1 + 0.8e^{T/600}}, \quad (3.90)$$

and [104]

$$v_D sat, h = \frac{1.5 \times 10^8}{1 + 20e^{T/100}}. \quad (3.91)$$

The drift velocity of the saturated carriers, electrons and holes, at different temperatures are depicted in Fig. 3.27.

3.2 Conclusions

An exploration of the increase in semiconductor conductivity at reduced temperatures is discussed in this chapter. Six primary scattering mechanisms which affect the semiconductor mobility are reviewed: phonon scattering, ionized impurity scattering, velocity saturation, carrier-to-carrier scattering, neutral impurity scattering, and surface scattering. The behavior of carriers in intrinsic and doped silicon, particularly due to changes in electron and hole populations as well as the dependence of the Fermi level on decreasing temperature, is discussed. Furthermore, insight into the electronic structure and energy distribution of semiconductors is provided, along with the effects of temperature variations on the effective mass of different bands. A foundational understanding of the complex interactions that determine semiconductor conductivity when operating in a cryogenic environment is offered in this chapter.

Chapter 4
Temperature-Frequency Boundary of Cryogenic Dynamic Logic

Cloud computing requires high performance, energy efficient processors and can support cryogenic operation since portability is not desired. Higher frequency circuits, required in cloud computing systems, can utilize dynamic logic circuits for improved performance. As compared to static logic circuits, dynamic circuits are faster, lower power, and require less area. Room temperature (RT, 300 K) dynamic logic circuits cannot however operate at low frequencies due to leakage currents and are highly sensitive to noise. Dynamic circuits at RT operate correctly only above certain frequencies since the state is temporarily stored on a capacitor. This drawback makes developing complex systems based on dynamic circuits challenging, since low frequency operation of these circuits is a difficult issue. Dynamic circuits operating at cryogenic temperatures bypass these issues [152].

Significant improvements in performance can be achieved by cooling the environment [35–37, 153]. Cryogenic electronics plays an important role in meteorology, quantum computing, high energy physics, and space-based technologies. Furthermore, applications which require a high signal-to-noise ratio and/or receive weak input signals benefit when operated at cryogenic temperatures [154]. Cryogenic environments also avoid thermal runaway in semiconductors [155]. Nevertheless, static logic is preferable at RT since these circuits are not affected by leakage currents. At cryogenic temperatures, alternatively, dynamic circuits are preferable due to the aforementioned speed, power, and area advantages [34]. One issue is at what temperature does dynamic logic outperform static logic. In this chapter, this temperature is determined for different frequencies. The key temperatures discussed in this chapter are liquid helium temperature (LHT) 4 K, liquid nitrogen temperature (LNT) 77 K, and room temperature 300 K. The relative performance of dynamic logic at these temperatures is also discussed.

The chapter is organized as follows: the characteristics of static and dynamic circuits are highlighted, and insight into MOSFET operation at cryogenic temperatures is summarized in Section 4.1. The cryogenic behavior of a MOSFET and a related HSpice model are discussed in Section 4.2. A dynamic circuit is characterized at

N. Zhuldassov, E. G. Friedman, *Cryogenic Microelectronic Systems for Ultra-Low Energy and Enhanced Performance*, https://doi.org/10.1007/978-3-031-90754-8_4

300 K and 4.2 K operation in Section 4.3. The temperature boundary at which dynamic logic should be used rather than static logic is discussed in Section 4.4. Some conclusions are offered in Section 4.5.

4.1 Background

Modern high complexity digital systems primarily deploy static logic. These systems can be enhanced by dynamic logic if operated at cryogenic temperatures. The temperature at which static logic should be replaced by dynamic logic depends upon the leakage currents; specifically, the subthreshold leakage current. The effects of subthreshold leakage current varies based on the style of logic circuit. Differences between static and dynamic logic are described in subsection 4.1.1. The behavior and advantages of dynamic circuits operating at LHT are discussed in subsection 4.1.2, and the behavior of a MOSFET operating at cryogenic environments is reviewed in subsection 4.1.3.

4.1.1 Static and Dynamic Logic

Many types of MOSFET logic have been developed since the 1970's. These circuits primarily fall into two categories, static logic and dynamic logic.

Logic circuits are combinatorial in nature and without memory. To reduce power consumption, static logic circuits can operate at lower frequencies or placed into a standby mode [152]. An example of a static logic circuit (a D flip flop) is shown in Fig. 4.1. In this master-slave flip flop, the state is maintained at the master stage output, and is transferred upon arrival of the next clock cycle to the slave stage.

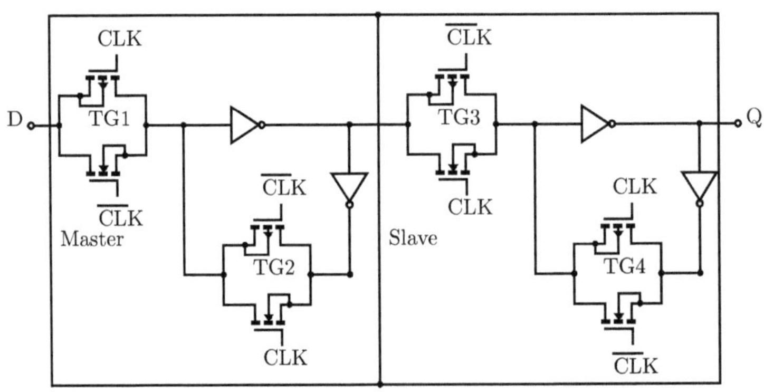

Fig. 4.1 Master-slave D flip flop as an example of a static logic circuit. Each state can be maintained indefinitely regardless of the clock frequency as long as power is maintained

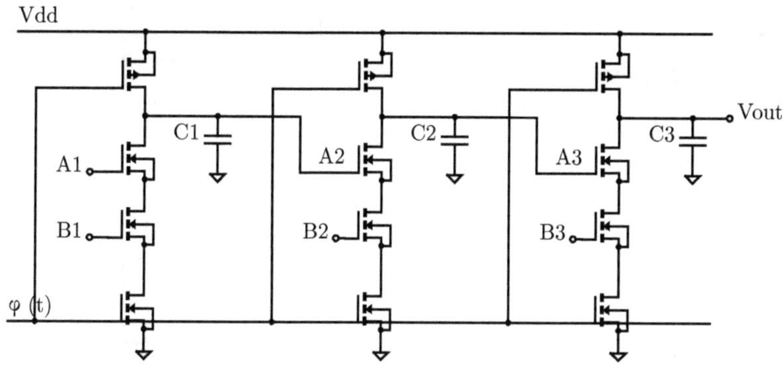

Fig. 4.2 Dynamic shift register. At low frequencies, the output degrades due to leakage currents

Dynamic logic circuits, alternatively, depend upon the temporary storage of information. Gate and stray capacitances store the charge in these circuits, maintaining the logic state. Dynamic circuits require a periodic refresh since the charge decays over time. In dynamic circuits, the logic and memory elements are the same circuit, and the charge is transferred between capacitors. This signal flow between capacitors is controlled by an external signal. An example of a dynamic logic circuit, specifically, a dynamic shift register, is shown in Fig. 4.2. The output and input of each stage of a shift register are connected to each other. At each transition of the clock signal, a bit of information is shifted by one position, passing in the input data and passing out the output data.

Dynamic logic requires less area and consumes less power than static circuits. Furthermore, dynamic circuits can operate at higher clock frequencies. These circuits, however, exhibit some undesirable characteristics. Dynamic logic circuits cannot operate at low frequencies due to the leakage of charge on the capacitors and exhibit lower noise immunity [156]. These issues cause another problem: the testability of a dynamic circuit is poor. Since it is not possible to maintain a logical state at DC in RT dynamic logic circuits, it is difficult to statically test a circuit and/or probe individual nodes [156].

4.1.2 Dynamic Logic

A dynamic logic circuit is shown in Fig. 4.3a. The capacitive load C_{out} maintains the output voltage V_{out}. The NMOS logic network behaves as a composite switch. As previously noted, V_{out} can decay within a short time due to leakage currents.

The clock signal $\varphi(t)$ controls the operation of the circuit, where the data flow from one stage to another stage across a cascaded network. The clock signal alternates between logic 0 and 1, described as the precharge and evaluate phases of the network. The precharge occurs at $\varphi = 0$. The PMOS transistor is active and

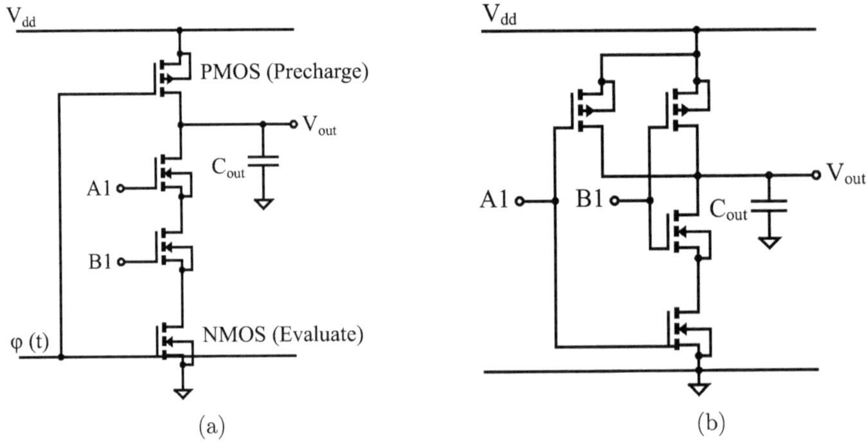

Fig. 4.3 Two input NAND gate, (a) dynamic logic, and (b) static logic

the NMOS transistor is cutoff; as a result, the output charges to V_{dd}. Although the output node is always charged to V_{dd}, the output does not always remain at logic 1. Rather, this phase pre-conditions the output, *i.e.*, the output node is charged to V_{dd} regardless of the input signals, preparing the node for the evaluation phase.

Conversely, at $\varphi = V_{dd}$, the NMOS transistor is active and the PMOS transistor is in cutoff during the evaluation process. The logic gate evaluates the inputs based on the signal polarity and establishes the output voltage. Two possible scenarios can occur. If the composite logic block forms a closed switch, the output capacitor discharges to ground, resulting in a logic 0 at the output. If, alternatively, the composite logic block is an open circuit, the charge is not discharged to ground, resulting in a logic 1 at the output. This state, however, cannot be maintained indefinitely due to leakage currents discharging the output capacitor. The charge at the output must therefore be read before the voltage drops below the logic 1 noise margin, introducing a temporal limit on the read time. Thus, these types of circuits are dynamic in nature [152].

These disadvantages of dynamic circuits no longer exist when operated at cryogenic temperatures. Since the leakage current at cryogenic temperatures is negligible, the charge on the capacitors does not leak as compared to room temperature operation. This property allows a dynamic circuit to operate at lower frequencies, essentially DC, which reduces the power consumption and supports low frequency, albeit cryogenic, testing.

At cryogenic temperatures, a MOSFET exhibits enhanced physical properties such as higher transient currents, negligible leakage currents, and increased subthreshold slope [157]. These properties are discussed in the following subsection.

4.1.3 MOSFET Characteristics Operating at Cryogenic Temperatures

Silicon MOSFETs operating at cryogenic temperatures exhibit significant improvements in performance and reliability as compared to room temperature operation. Since the late 1980's, cryogenic MOSFETs have, until recently, received little attention. Early studies on the behavior of MOSFETs at cold temperatures are described in [158–164]. The topic has once again become important due to the growing interest in cloud computing, enhancing the need for energy efficient, stationary computing platforms.

The carriers within the substrate of a CMOS transistor start to freeze out at 77 K [105]. With a voltage applied between the source and drain, the carriers, driven by energy from an applied electric field, form an inversion layer, becoming conductive. The mobility of the channel increases at lower temperatures due to less carrier scattering caused by lattice vibrations, but is limited by Coulombic scattering at lower gate voltages and surface roughness at higher gate voltages [165]. As a result, the speed of MOSFET devices operating at cryogenic temperatures is greater than at room temperature. For example, a 62% improvement in speed has been observed in a ring oscillator operating at 4 K [105]. A MOSFET operating at cryogenic temperatures exhibits the following advantages as compared to 300 K behavior: negligible leakage currents, higher subthreshold slope, insignificant electromigration, fewer thermal induced failures such as oxide degradation, higher transconductance, and no latch-up [105].

4.2 CMOS Characterization and Modeling at LHT

The design of high performance circuits requires efficient simulation models. A model of a CMOS transistor operating at cryogenic temperatures is crucial for circuit analysis. The properties of LHT operation and the development of an HSpice model are described, respectively, in subsections 4.2.1 and 4.2.2.

4.2.1 Properties of CMOS Technology at 4.2 K

The MOSFET considered in this chapter is a 160 nm CMOS transistor [166]. The transistor has a width of 2.32 μm and a channel length of 160 nm with a thin oxide. The drain current I_D at 4.2 K is 40% higher than the drain current at room temperature due to the increase in the carrier mobility (μ). The threshold voltage is highly temperature dependent due to the carrier freezeout effect and subsequent changes in the Fermi potential [167]. As compared to room temperature operation, the threshold voltage increases from 0.55 volts to 0.7 volts and the subthreshold

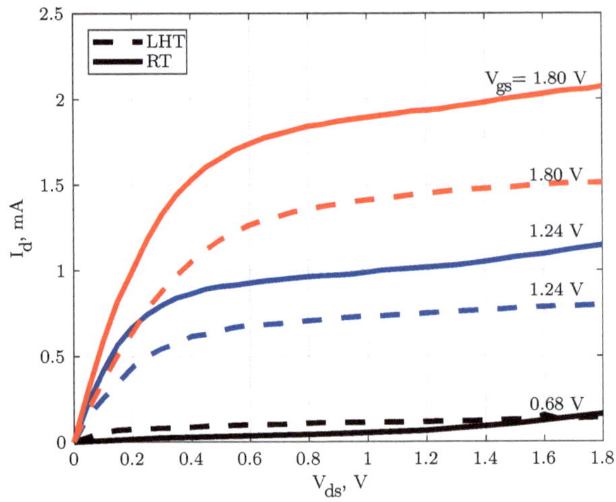

Fig. 4.4 Experimental $I_D(V_{DS})$ characteristics of 160 nm CMOS for different V_{GS} [166]

slope (SS) is 3.8 times steeper in LHT than RT, increasing from 87.0 mV/dec to 22.8 mV/dec. The kink effect is observed in thick oxide CMOS at LHT [166]. Since the kink effect is not observed in thin oxides, which is assumed here, the kink effect is not considered. I-V curves for different V_{GS} at room temperature (300 K) and LHT (4.2 K) are shown in Fig. 4.4 [166].

The SS plays a significant role in leakage current, greatly lowering the leakage current at lower temperatures. The drain current in the subthreshold regime can be modeled [157] as

$$I_{DS} \approx I_S \mathrm{e}^{\frac{q(V_{GS}-V_T)}{nkT}} \left(1 - \mathrm{e}^{-\frac{qV_{DS}}{kT}}\right), \tag{4.1}$$

where I_S is the saturation current, q is the electron charge, k is the Boltzmann constant, and T is the absolute temperature. The subthreshold slope factor n [166] is

$$n(T) = 1 + \frac{C_{dep}(T)}{C_{ox}}. \tag{4.2}$$

$C_{dep}(T)$ is the channel and bulk depletion capacitance, while C_{ox} is the gate oxide capacitance. Note that I_{DS} is exponentially dependent on the threshold voltage, temperature, and subthreshold slope factor.

An expression for the subthreshold slope [166] is

$$SS(T) = \left[\frac{\partial \log I_D}{\partial V_{GS}}\right] = \ln(10)\frac{nkT}{q}. \tag{4.3}$$

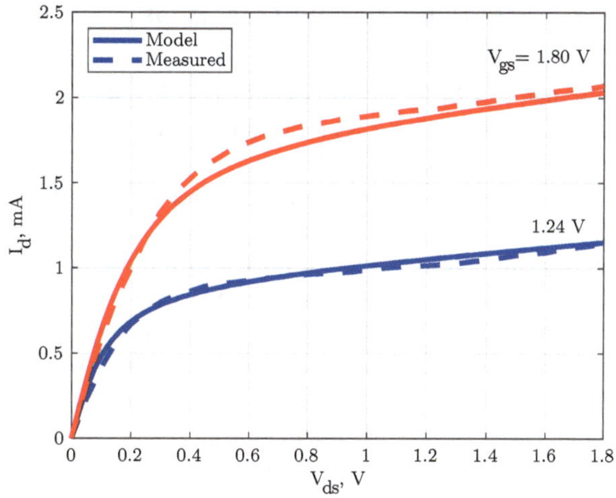

Fig. 4.5 I_D vs V_{DS} for $V_{GS} = 1.24$ volts and 1.8 volts for measured and simulated CMOS. An average error below 4% and maximum error below 5% between the model and the experimental data are observed

At cryogenic temperatures, however, due to incomplete ionization of the dopants, carrier freezeout occurs [105]. This effect causes the SS to no longer be proportional to the temperature at LHT.

4.2.2 *HSPICE Model*

The Phillips MOS11 HSpice model is used for the 160 nm CMOS transistor [166]. The temperature dependent parameters have been modified to fit the experimental data. The revised parameters are based on [166].

To match the properties of the device to a device operating at 4.2 K, parameters such as VFB BETSQ, THESRR, THESATR, SDIBLO, ALPR and KOR are fitted to the LHT data. Certain fitting parameters for the original temperature of the MOS11 model are inaccurate at LHT, and therefore defaulted to zero. The I-V curves exhibit an average error below 4% and a maximum error below 5% for V_{GS} above the threshold voltage. I_D versus V_{DS} for this model and measured data at 4.2 K are shown in Fig. 4.5.

4.3 Dynamic Circuit Characterization and Simulation

The behavior of dynamic logic circuits operating at 300 K and 4.2 K is described in this section, and the results are confirmed with HSpice simulations. The circuit is presented in subsection 4.3.1, and the behavior of this circuit at RT and LHT is described in subsection 4.3.2. Results of the simulation at room temperature and cryogenic temperature are shown, respectively, in subsections 4.3.3 and 4.3.4.

4.3.1 Dynamic Circuit Operation

A dynamic CMOS circuit, as shown in Fig. 4.3(a), operating at room temperature (300 K) and LHT (4.2 K) is considered here. The circuit consists of four transistors and a 20 fF capacitive load.

In the dynamic logic block, $A1$ is set to 1 and $B1$ is set to 0. During precharge, the voltage across the output is charged to logic 1. However, during the evaluation phase, although the NMOS transistors are on, the output remains at logic 1 since the logic block behaves like an AND gate. Although $A1$ is 1, $B1$ is 0, therefore V_{out} is not connected to ground. The clock signal and output voltage are shown in Fig. 4.6(a), with operation at room temperature and a frequency of 10 MHz. During the evaluation phase, the output voltage degrades due to leakage currents. Since the frequency is sufficiently high, the voltage does not drop below the threshold voltage (which is 0.55 volts at RT, labeled in the figure as logic 1).

4.3.2 Dynamic Circuit Behavior at 300 K and 4.2 K

To develop an expression for the decay time, the subthreshold current described by (4.1) is equated to the capacitor current,

$$I_S e^{\frac{-qV_T}{nkT}} \left(1 - e^{-\frac{qV_{out}}{kT}}\right) = C_{out} \frac{\partial V_{out}}{\partial t}. \tag{4.4}$$

The decay time t of capacitor C_{out} is

$$t = \frac{C_{out} kT}{I_S q} e^{\frac{qV_T}{nkT}} \ln \left[e^{\frac{q}{kT}} (V_{dd} - V_{out}) - 1 \right]. \tag{4.5}$$

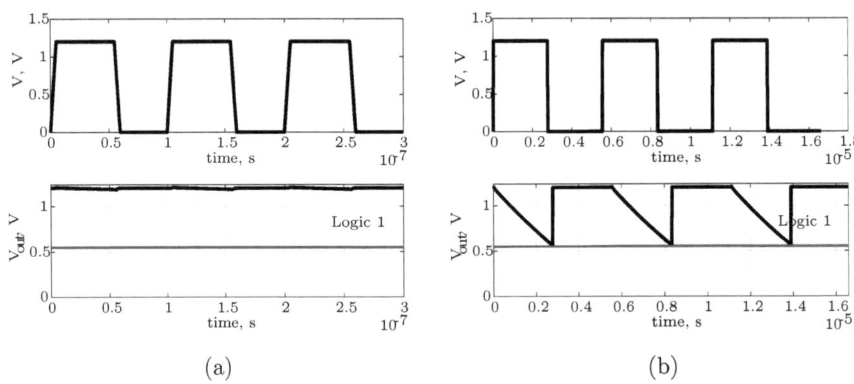

Fig. 4.6 Dynamic circuit at RT operating at (a) 10 MHz, and (b) 180 kHz. 10 MHz operation shows the circuit operating normally, while 180 kHz is the minimum functional frequency

The decay time of V_{out} for $V_T = 0.55$ volts and $T = 300$ K is 3.1 μs. The minimum frequency for proper operation at room temperature is 161 kHz.

For cryogenic LHT operation, the charge decay time is significantly longer than the age of the Universe. A dynamic circuit at LHT can therefore operate at DC.

4.3.3 Simulation of Dynamic Logic Circuit at 300 K

For room temperature operation, the minimum frequency is 180 kHz. The output voltage at this frequency is shown in Fig. 4.6(b). During the evaluation phase, the charge on the output capacitor drops below the threshold voltage; in the next stage, this voltage is no longer at logic 1. Since 180 kHz represents a decay time of 2.78 μs, the difference between the simulation and analytic solution for the decay time is 10%.

If the frequency drops further, the charge completely leaks away before arrival of the next clock pulse when operating at RT. An example of this condition is shown in Fig. 4.7(a), where the frequency is 500 Hz. Note that the charge on the output capacitor completely leaks away during the evaluation phase.

4.3.4 Simulation of Dynamic Logic Circuit at 4.2 K

In comparison to the dynamic circuit operating at RT and 500 Hz, the same circuit at LHT exhibits no drop in voltage at 500 Hz, as shown in Fig. 4.7(b). This result agrees with the analytic solution. Note that the frequency can be further reduced, allowing the dynamic circuit at 4.2 K to operate at DC.

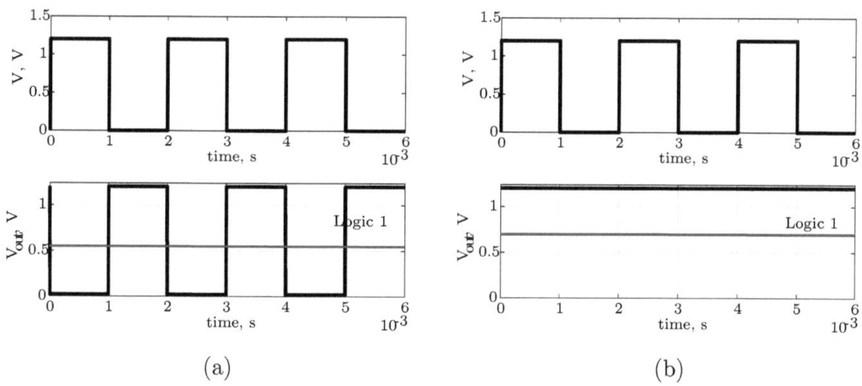

Fig. 4.7 Dynamic circuit operating at 500 Hz in (a) RT, and (b) LHT. The circuit shows an incorrect output at RT, while at LHT the circuit operates properly

The maximum speed and power at room temperature and liquid helium temperature are compared. The highest frequency at RT is 2.5 GHz while dissipating 37.1 nW. The frequency and power consumption of the same circuit at LHT is, respectively, 20 GHz and 7.2 fW, eight times faster and 5x10^6 times less power. These results demonstrate the speed and power advantages of LHT operation.

4.4 Temperature Breakpoint of Static and Dynamic Logic

The operating temperature at which dynamic logic (Fig. 4.3(a)) is preferable to static logic (Fig. 4.3(b)) is discussed here as the breakpoint (or boundary) temperature. The range of temperatures being considered is between 4.2 K and 300 K. The simulation setup is presented in subsection 4.4.1. Operation of dynamic circuits at LHT, LNT, RT, and at certain transitional temperatures are described in subsection 4.4.2. The breakpoint temperatures are also described in subsection 4.4.2 at different frequencies.

4.4.1 Simulation Setup for Evaluating Dynamic Logic

A two input dynamic NAND gate is shown in Fig. 4.3(a). The circuit consists of four transistors and a capacitive load, $C_{out} = 0.1 fF$ [168]. The voltage supply V_{DD} is 1 volt. The NMOS logic network behaves as a composite switch. The output voltage V_{out} is stored on a capacitive load C_{out}. As previously noted, at RT, V_{out} decays due to leakage currents. Operation of a dynamic NAND gate where the leakage current is a significant issue is exemplified by the following case. In a dynamic NAND gate during the precharge phase (see Fig. 4.3(a)), $A1$ is set to 1 and $B1$ is set to 0, while the voltage across the output charges to logic 1. During the evaluation phase, although some NMOS transistors are on, the output remains at logic 1 since a signal is only transmitted if both $A1$ and $B1$ are on. The clock signal and output voltage at RT operating at a frequency of 20 GHz are shown in Fig. 4.8(a). During the evaluation phase, the output voltage degrades due to leakage currents. If the frequency is sufficiently high, the voltage does not drop below the noise margin high (which, in this example, is 0.65 volts at RT, labeled in Figure 4.8 as logic 1).

A 160 nm CMOS transistor operating at cryogenic temperatures is used [34]. For smaller technology nodes, the precise breakpoint temperature changes. However, as the temperature dependent characteristics of CMOS devices experience similar changes at cryogenic temperatures [157], the cryogenic behavior of the 160 nm CMOS technology model [34] used here can be extended to more deeply scaled, thin oxide CMOS technologies.

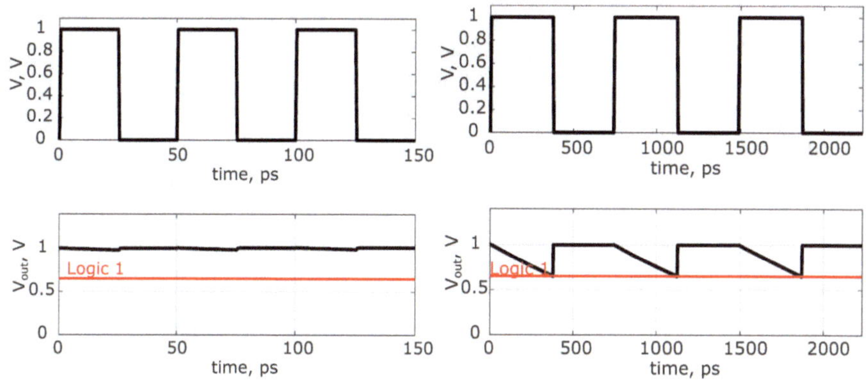

(a) (b)

Fig. 4.8 Dynamic circuit operating at RT, (a) 20 GHz, and (b) 1.344 GHz. Note that the circuit operates normally at 20 GHz while 1.344 GHz is the minimum operational frequency

4.4.2 Temperature Breakpoint

Dynamic circuit operation at a specific temperature depends upon the frequency of operation. If the dynamic logic cannot operate at a specific temperature and frequency due to leakage currents, static logic is preferred. This temperature-frequency boundary where the output voltage of the dynamic logic drops below the noise margin high is the breakpoint temperature. To determine this breakpoint temperature, (4.5) is evaluated at different temperatures and frequencies.

Based on (4.5), the decay time of V_{out} for $V_T = 0.25$ volts and $T = 300$ K is 0.414 ns. The minimum frequency to correctly operate at room temperature is analytically determined to be 1.209 GHz. In simulation, the minimum frequency of operation is 1.344 GHz, within 11% of the analytically determined frequency. The simulated output voltage at this frequency is shown in Fig. 4.8(b). During the evaluation phase, the charge on the output capacitor drops below the threshold voltage. In the next stage, this voltage is no longer at logic 1. 1.344 GHz represents a decay time of 0.372 ns. The difference between the simulation and analytic solutions for the decay time is 10%. These delay times are listed in Table 4.1.

Table 4.1 Operating frequency at different temperatures

	Minimum Frequency		
Temperature	**(Simulated)**	**(Analytic)**	**% Error**
300 K (RT)	1.209 GHz	1.344 GHz	10.0%
90 K	49.9 MHz	56.8 MHz	12.1%
77 K (LNT)	29.7 MHz	33.6 MHz	11.6%
4.2 K (LHT)	DC	DC	—

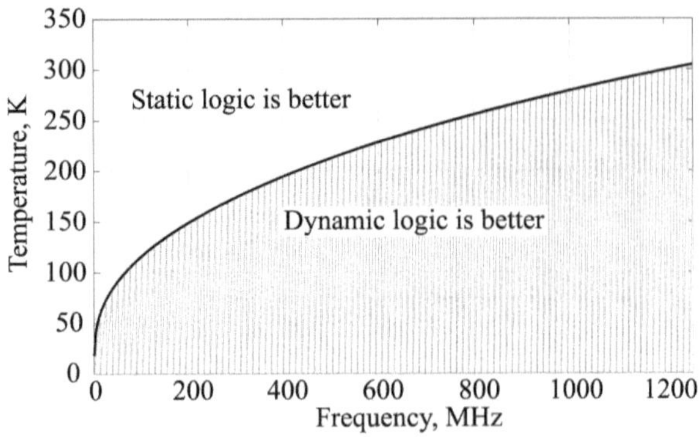

Fig. 4.9 Breakpoint temperature at which dynamic logic is preferable to static logic at different frequencies, 160 nm CMOS technology

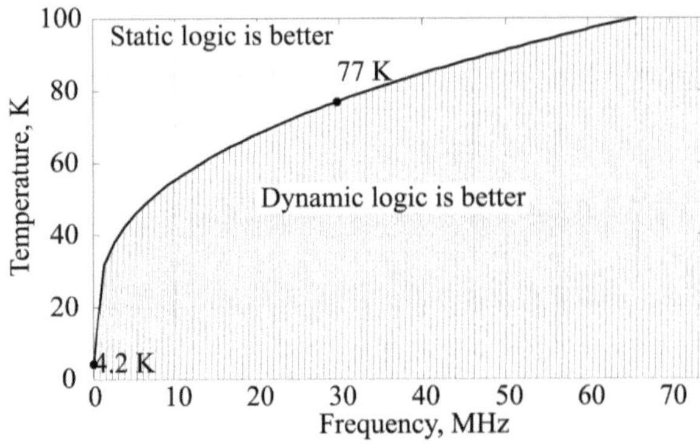

Fig. 4.10 Breakpoint temperature below 100 K at which dynamic logic is preferable to static logic at different frequencies

The breakpoint temperature at different frequencies is shown in Fig. 4.9. The breakpoint temperature-frequency characteristic is exponential, as described by (4.5). A scaled version of the temperature-frequency characteristic at temperatures below 100 K is shown in Fig. 4.10.

As noted in Fig. 4.9, when operating at RT at frequencies below 1.344 GHz, the charge leaks away before arrival of the following clock pulse. At these lower frequencies, static logic is necessary. At lower temperatures, the frequency at which dynamic logic functions correctly can be decreased. As the temperature approaches liquid nitrogen, the breakpoint curve becomes steep due to the exponential nature of the subthreshold leakage current. At 77 K, dynamic logic should be used when operating at frequencies above 29.7 MHz.

Note that the frequency can be further reduced at lower temperatures, allowing dynamic logic at 4.2 K to operate at DC. An example of this condition is shown in Fig. 4.11. The circuit at LHT exhibits no drop in voltage at 50 MHz, agreeing with the analytic solution. Dynamic logic is therefore preferable at LHT.

DC operation, however, is available at temperatures higher than LHT. The charge leakage time at different temperatures is shown in Fig. 4.12. The concept of application age is introduced here, where the age of the application indicates the time during which a circuit has to retain sufficient charge to maintain the logic state.

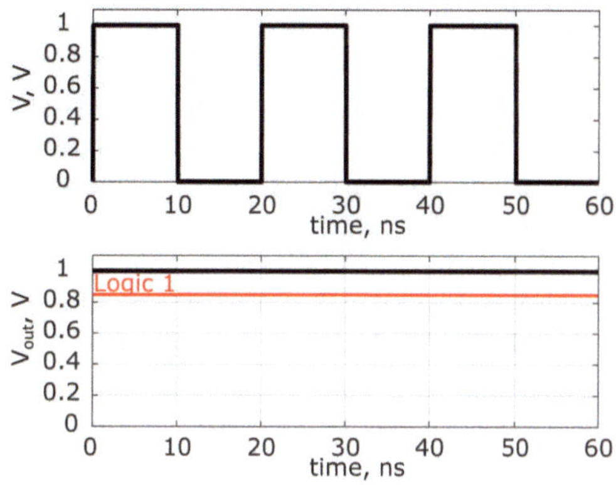

Fig. 4.11 Dynamic circuit operating at 50 MHz at LHT. The circuit operates properly at LHT

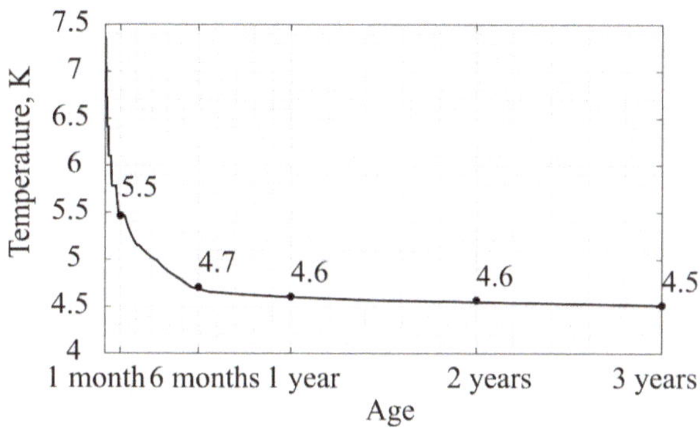

Fig. 4.12 Age of application at different temperatures to maintain DC operation

As shown in Fig. 4.12, the application age is one month at 5.5 K, one year at 4.6 K, and three years at 4.5 K. Therefore, if system operation requires an age of three years, the circuit should be operated at or below 4.5 K.

Summarizing, as the frequency increases from 1 Hz to 1.209 GHz, the breakpoint temperature increases, respectively, from 11 K to 300 K, as shown in Fig. 4.9. Therefore, if the circuit operates below 1.209 GHz at a temperature above 300 K, static logic should be used. Dynamic logic can operate at 1 Hz at temperatures below 11 K and at DC below 4.5 K, as illustrated in Figs. 4.10 and 4.12.

4.5 Conclusions

Interest in cloud computing has been growing, demanding greater energy efficiency. Due to the stationary nature of these platforms, cloud computing can support cryogenic operation. This characteristic suggests the use of dynamic logic in cloud computing centers operating at cryogenic temperatures to support high frequencies while dissipating low power. The primary objective of this chapter is to clarify at which temperature and frequency dynamic logic is preferable to static logic.

Dynamic logic operating at temperatures ranging from liquid helium to room temperature is considered here. The performance of dynamic circuits operating at the transitional temperature boundary between room temperature and liquid helium temperature is evaluated at several frequencies. At liquid helium temperature, dynamic circuits can operate at DC. Dynamic logic is therefore preferable over static logic at any frequency if operated at LHT. At higher temperatures, the time during which the charge leaks away is exponentially less. Furthermore, if DC operation requires holding the charge for one month or less, the temperature can be increased from LHT to 5.5 K. Dynamic logic is preferable at temperatures below 11 K when operated above 1 Hz. At liquid nitrogen temperature, dynamic logic is preferable when operating above 29.7 MHz. Above 1.209 GHz, dynamic logic can be used at any temperature.

Chapter 5
Thermal Optimization of Hybrid Systems

The demand for high performance computing (HPC) has greatly increased over the past several decades, driven by the rise in computationally intensive, large scale applications, particularly cloud computing. Further advancements in HPC systems require overcoming a large number of challenges, including energy efficiency, thermal management, and system performance. The energy consumption of a typical data center ranges from tens to hundreds of megawatts [169]. The annual global energy consumption for HPC is estimated at 200 TWh, and is expected to increase fourfold by 2030 [49]. Qualitatively different computational technologies are necessary to sustain this rapid growth in computing. Cryogenic technologies can potentially reduce the power consumption of large scale, stationary computing systems by several orders of magnitude, including the energy cost of the refrigeration [36, 45]. The cooling capacity at 4 K is however often insufficient to efficiently dissipate the heat generated by the circuitry [65]. Furthermore, as illustrated in Fig. 5.1, it may be advantageous to place certain circuits at lower temperatures while, as the temperature is reduced, other circuits are placed at higher temperatures.

The operating temperature also greatly affects the architecture of a heterogeneous computing system. By adjusting the operating temperature of each subsystem, the performance and power of the overall system can be better controlled. For example, the temperature of a cryogenic CMOS subsystem can be increased; different technologies can be placed at different stages of a cryocooler to reduce refrigeration costs. The latency and power dissipation of this subsystem however may also increase. Furthermore, refrigeration of nearby subsystems operating at a lower temperature can be affected if these subsystems are not thermally isolated.

An approach where different technologies are placed at different stages of the refrigerator has previously been proposed [174, 175]. A hybrid temperature system exploits multiple stages of a cryocooler; in [174], a Sumitomo SRDK-101DP-11C cryocooler with 4 K and 60 K stages is introduced. Low temperature superconductive circuits are located at the 4 K stage, higher temperature semiconductor circuits,

N. Zhuldassov, E. G. Friedman, *Cryogenic Microelectronic Systems for Ultra-Low Energy and Enhanced Performance*, https://doi.org/10.1007/978-3-031-90754-8_5

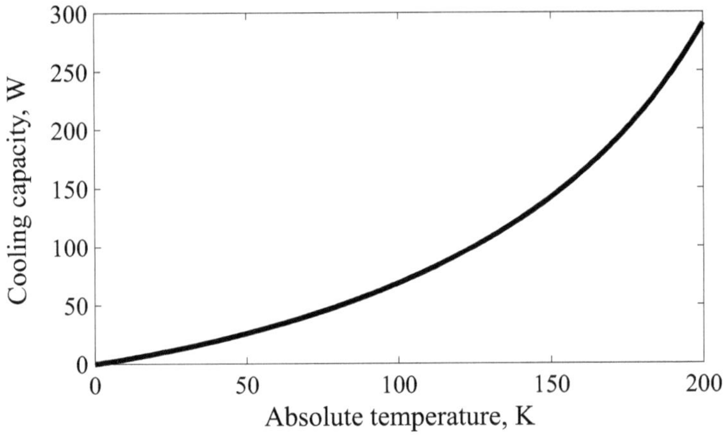

Fig. 5.1 Available cooling power per 1 kW input power. The data is based on the technical specifications of commercially available cryogenic coolers [170–173]

such as analog filters and low noise amplifiers, are placed at the 60 K stage, and the remaining electronics are placed at room temperature. These studies utilize different stages within a cryocooler, but do not consider the possible range of temperatures within a specific stage. For example, the second stage of the Sumitomo cryocooler in [174] is set to 60 K, while the available temperature range of the second stage of this cryocooler can vary between 60 K to 80 K.

This range of available temperatures of each stage within a cryocooler is exploited here to enhance the overall performance of computing systems under a target heat load constraint. A methodology for optimizing the temperature of each component within a cryogenic system is proposed. The total power consumed by the system is minimized while maintaining acceptable performance. The methodology is confirmed in a case study requiring cryogenic operation, a quantum computer, a technology which potentially will accelerate a wide range of computing tasks, such as prime factorization, quantum simulation, and complex optimization [176, 177].

The chapter is organized as follows: in Section 5.1 the problem is formulated, the thermal behavior of the system is discussed, and insight into the organization of cryogenic computing systems are described. A proposed methodology is described in Section 5.2. An example case study, a hybrid quantum computing system, optimized using the proposed methodology, is presented in Section 5.3. Some conclusions are offered in Section 5.4.

5.1 Background

For efficient integration of cryogenic computing systems, electronic circuits operating at cryogenic temperatures are necessary [65]. The design objective is to

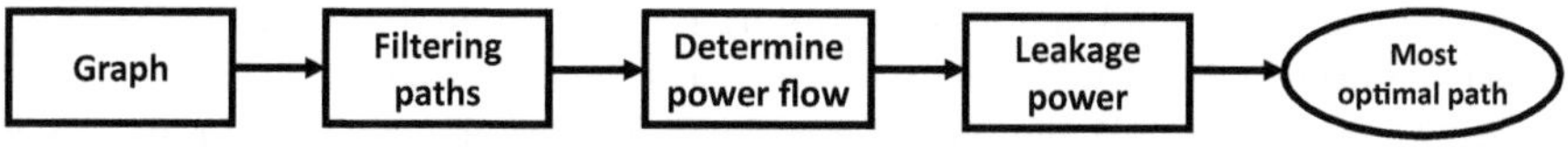

Fig. 5.2 Flowchart of the proposed methodology. A graph of the system is initially generated. Paths connecting the initial and final state of the process satisfying the constraints are determined. The power flow within the system is subsequently evaluated to determine the total power consumed by the system. The optimal system is based on the total power consumption and delay of each path

determine a set of temperatures for each of the components at which the total power consumption or delay is minimized while satisfying target constraints, which denotes optimal operation of a cryogenic system.

A methodology is proposed to determine the optimal temperature of the different parts of an electronic system. Four steps are performed in the methodology, as shown in Fig. 5.2. A graph of the system is initially formulated, and the available range of temperatures for each component within the system is determined. An algorithm to evaluate the set of optimal temperatures, exploiting graph theory [178], is proposed. After determining the set of temperatures which satisfies the constraint, a thermal model of the system is generated to evaluate the flow of heat (or power) from unit to unit. The heat flow depends upon the thermal conductance between units. The rate of heat flow depends upon the temperature of the connecting wires. Furthermore, the heat flow is used to estimate the leakage power; specifically, the power lost from the additional cooling required at lower temperatures due to the flow of heat from higher temperature components. The net power consumption at a specific set of temperatures therefore includes the leakage power between temperature zones. Optimal operation of the system, considering delay and power constraints and the heat flow among the components, sets the temperature for each component.

The rest of the section is organized as follows. The problem formulation based on graph theory is described in Section 5.1.1. A thermal model of the system is discussed in Section 5.1.2. A hybrid quantum computing system as a case study is described in Section 5.1.3.

5.1.1 Formulation of Thermal Optimization Problem

The objective is to determine a suitable operating temperature at each step of the process. Temperature optimization of a process can be described as a directed acyclic multiweighted multigraph $G := \langle S, U, W \rangle$. A finite set of states in the process $S = \{S_1, S_2, \ldots, S_n\}$ specifies an instance of the temperature optimization problem. A set of edges is denoted by U and represents a unit performing a computational step. Parallel edges correspond to computing unit i which comprise a subset $U_i \subseteq U$. A typical refrigeration system operates at a specific set of temperatures, such as liquid helium temperature (LHT) or liquid nitrogen temperature (LNT) in cryogenic CMOS. Index j represents the set of available temperatures, $T = \{T_1, T_2, \ldots, T_j\}$. A unit at different temperatures at each step is represented by $u_{i,j} \in U_i$. Two weights are associated with each edge $u_{i,j}$, $W := \langle p, d \rangle \in \mathbb{R}^2_{>0}$, where p and d represent, respectively, the power consumption and delay of a unit at a specific temperature.

A set of operating temperatures corresponding to each computing unit constitutes a path connecting the source to the sink of the process graph. Path π is the collection of specific edges between two endpoints of a process,

$$\pi = \big(U_1(T_j), U_2(T_j), \ldots, U_i(T_j)\big) . \tag{5.1}$$

The power consumption of a process is the sum of the power weights along a path, $P(\pi) = p_1 + p_2 + \cdots + p_{n-1}$. The weight of an edge represents the power consumption of a unit. Similarly, the delay of the process is the total cost of the weights, which represents the delay of a unit of the edges along a path, $D(\pi) = d_1 + d_2 + \cdots + d_{n-1}$. Given set U at different temperatures performing a computation among states S, the temperature optimization problem is to determine a path connecting the source and sink states of a system that minimizes the total power $P(\pi)$ while constraining the total delay of the system, $D(\pi)$.

$$\text{Minimize: } P(\pi), \tag{5.2}$$

$$\text{subject to: } D(\pi) \leq D_{max}. \tag{5.3}$$

An example of the process containing three units and four states is depicted in Fig. 5.3. Each unit can operate at three different temperatures, as denoted by the parallel edges between adjacent states. A power and delay are associated with each edge. A path $\pi = (u_{1,1}, u_{2,3}, u_{3,2})$, highlighted in bold, corresponds to computing units u_1, u_2, and u_3 operating at, respectively, temperature T_1, T_3, and T_2. The total power consumption of the highlighted path is $P(\pi) = p_{1,1} + p_{2,3} + p_{3,2}$. The delay of the highlighted path is $D(\pi) = d_{1,1} + d_{2,3} + d_{3,2} \leq D_{max}$.

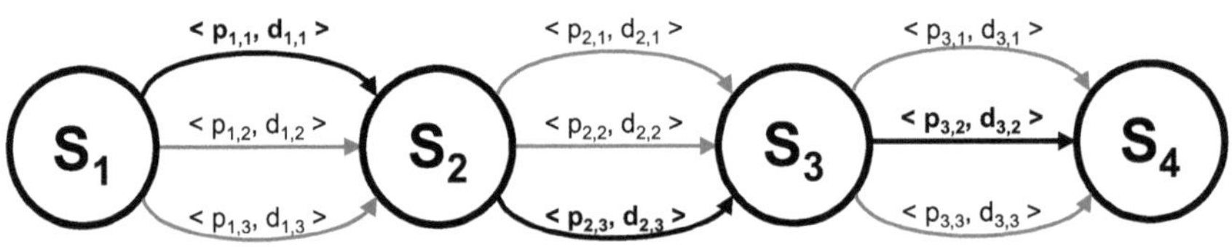

Fig. 5.3 Example of the temperature design process. The edges between two states describes a unit at different temperatures. An example of a path is shown highlighted in bold. The power consumption of this path is $P(\pi) = p_{1,1} + p_{2,3} + p_{3,2}$. The delay of the path is $D(\pi) = d_{1,1} + d_{2,3} + d_{3,2}$

The problem of finding the optimal set of temperatures resembles the knapsack problem [179] or multiple knapsack problem [180]. The optimization problem described here is however different. In the knapsack problem, the set of items maximizing the total value is determined while not exceeding the weight limit. In the multiple knapsack problem, the value of the items in two or more knapsacks is considered. The primary difference is related to the type of units stored in each bag. Since each unit has a unique set of power and delay weights, the knapsack analogy of this problem can be described as a multiple knapsack problem with a unique set of items stored in each knapsack. The cumulative value of each knapsack is maximized while constraining the total weight.

5.1.2 Thermal Model

Since computing units are not ideally isolated from each other, any temperature difference between units produces heat flow. The transfer of heat between computing units significantly contributes to the heat load of the cooling system. The choice of path contributes to this transferred power. A refrigerator requires different power at different temperatures due to the difference between the ambient and operating temperatures [181]. Similarly, a system with a significant difference in temperature between adjacent units likely requires greater refrigeration power as compared to a system with a smaller temperature difference [181]. Newton's law of cooling notes that the rate of temperature change of a body depends upon the ambient temperature and body temperature, and suggests that additional cooling power is required due to heat leaking from a higher temperature unit to a lower temperature unit. The net cooling power therefore not only depends upon the cooling power of each individual unit, but also depends upon the heat flow between units.

The thermal behavior of a system can be efficiently described based on an analogy with electrical circuits. This analogy between electrical and thermal processes is summarized in Table 5.1. While electric charge q is transported across a conductor in an electrical circuit, heat Q_T is transported through a thermal conductor. The heat flow (or power) per unit time q_T is analogous to the flow of current i, the electric charge per unit time. Similar to the potential difference ΔV driving an electric current in a conductor, heat is driven by the difference in temperature ΔT.

Table 5.1 Thermal-electric analogy

Thermal parameter	Symbol	Electrical parameter	Symbol
Heat [Joule]	Q_T	Charge [Coulomb]	q
Heat flow [Watt]	q_T	Current [Ampere]	i
Thermal capacitance [J/K]	C_T	Capacitance [Farad]	C
Thermal resistance [K/W]	R_T	Resistance [Ω]	R
Temperature [K]	T	Voltage [Volt]	V

By applying Fourier's law [182], the thermal resistance between units is

$$R_T = \frac{\Delta T}{q_T}. \tag{5.4}$$

Heat flow within a system is determined for each set of temperatures based on a thermal model. An example of a system containing six CPUs is shown in Fig. 5.4(a). The CPUs are represented by a thermal capacitance and are labeled as C_n, where n is the CPU number. Similar to a voltage source, the refrigerators are the heat source for the thermal capacitors and represent the voltage source in a thermal circuit. Since a refrigerator cannot instantly regulate the temperature of a CPU, the refrigerators are connected through thermal resistance R_r. Similarly, the CPUs are connected by thermal resistances. In this example, three types of connections among the CPUs exist, corresponding to three thermal resistances, R_1, R_2, and R_3. The connection between the adjacent units is R_1, diagonally adjacent units are connected through thermal resistance R_2, and the relationship between the units separated by one other unit is represented by R_3. The temperature of each unit in this example is maintained at T_i. The flow of power (or heat) between units is based on an estimate of the currents within the thermal circuit.

Since CPUs are often located in different refrigerators (or chambers), the thermal conduction between the CPU and refrigerator is much higher than the conduction with another CPU (since $R_r \ll R_1, R_2, R_3$). The thermal resistance between the refrigerator and CPUs (R_r) is therefore assumed to be negligible. The thermal circuit of the system shown in Fig. 5.4(a) can therefore be simplified to Fig. 5.4(b).

The flow of heat within the system is described by a system of linear expressions. A thermal resistance grid between the units is initially described as

$$R_T = \begin{matrix} & \begin{matrix} U_1 & U_k & U_n \end{matrix} \\ \begin{matrix} U_1 \\ U_k \\ U_n \end{matrix} & \begin{bmatrix} R_{1,1} & \cdots & R_{1,n} \\ \vdots & \ddots & \vdots \\ R_{n,1} & \cdots & R_{n,n} \end{bmatrix} \end{matrix}, \tag{5.5}$$

where the thermal resistance between units U_k and U_n is represented by $R_{k,n}$. If the units are thermally isolated, i.e., the units are not thermally connected, the thermal resistance is assumed infinite. The thermal resistances across the diagonal of the matrix are also equated to infinity, as each diagonal element represents a relationship of each unit to itself. Since the temperature of each unit is extracted from a graph within the proposed algorithm, the matrix of temperature differences (or thermal potential differences) can be established as

$$\Delta T = \begin{matrix} & \begin{matrix} U_1 & U_k & U_n \end{matrix} \\ \begin{matrix} U_1 \\ U_k \\ U_n \end{matrix} & \begin{bmatrix} \Delta T_{1,1} & \cdots & \Delta T_{1,n} \\ \vdots & \ddots & \vdots \\ \Delta T_{n,1} & \cdots & \Delta T_{n,n} \end{bmatrix} \end{matrix}. \tag{5.6}$$

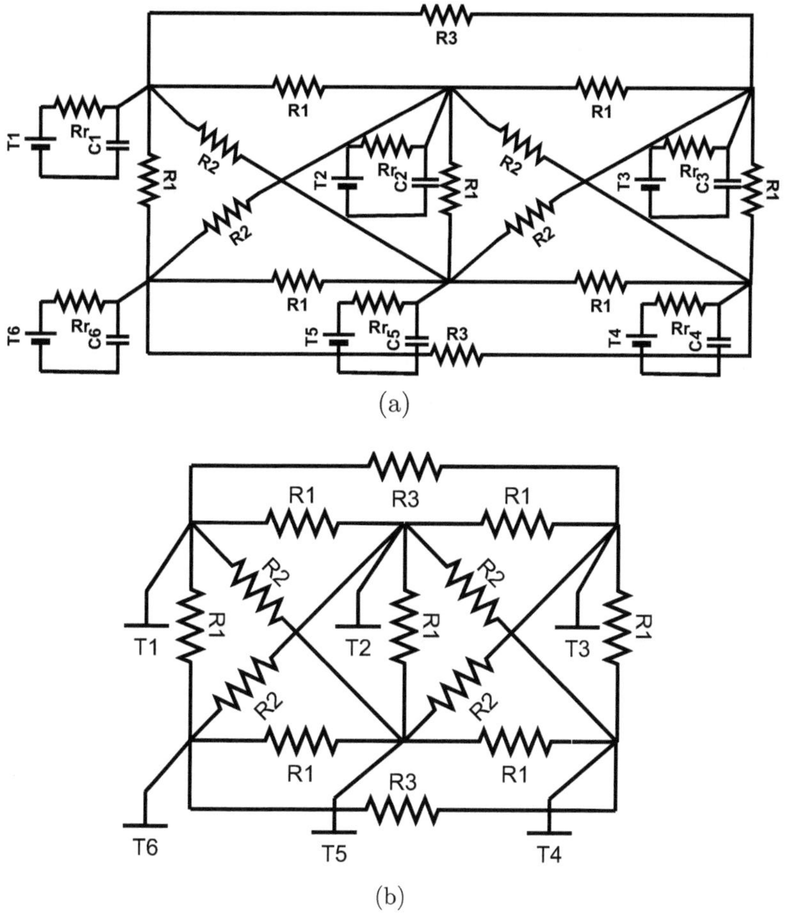

Fig. 5.4 Thermal model of a system with six CPUs, (a) complex model, and (b) simplified model. The complex model includes four types of temperature interactions: R_1, R_2, R_3, and refrigerator resistance R_r. The simplified thermal model includes only three types of temperature interactions: R_1, R_2, and R_3

Each element, representing a temperature difference between units, corresponds to an index. For example, $\Delta T_{k,n}$ represents the temperature difference between units k and n. As noted, the diagonal elements are each equal to zero. The power flow q_T between each unit can therefore be evaluated by elementwise matrix division using the electrical analogy, $q_T = \frac{\Delta T}{R_T}$, as

$$
q_T = \begin{matrix} \\ U_1 \\ U_k \\ U_n \end{matrix} \begin{matrix} U_1 \quad U_k \quad U_n \\ \begin{bmatrix} \frac{\Delta T_{1,1}}{R_{1,1}} & \cdots & \frac{\Delta T_{1,n}}{R_{1,n}} \\ \vdots & \ddots & \vdots \\ \frac{\Delta T_{n,1}}{R_{n,1}} & \cdots & \frac{\Delta T_{n,n}}{R_{n,n}} \end{bmatrix} \end{matrix} . \tag{5.7}
$$

The power flowing to or from each unit can be evaluated by summing all of the elements along each row as

$$\Delta P = q_T \mathbf{1}_n, \tag{5.8}$$

where

$$\mathbf{1}_n = [1, \ldots, 1]^\mathsf{T}. \tag{5.9}$$

Since the thermal resistance of a material varies depending upon the absolute temperature, the thermal resistance can be adjusted to more accurately characterize the heat flow between units [183, 184]. The thermal conductivity of cryogenic cables, made from materials such as stainless steel, niobium-titanium alloy, and beryllium copper, exhibits a rising trend with increasing temperature [185–187]. The thermal conductivity of beryllium copper can be linearly approximated [187], whereas that of stainless steel can be represented by a dual-line approximation [186].

5.1.3 Control of Quantum Computing Systems

Qubit control requires readout and generation of electronic and optical signals [65]. The readout measures the resonant frequency of a resonator in the case of transmons [67] or the impedance of a charge detector in the case of spin qubits [68]. These measurements are used to evaluate extremely low noise signals, as a quantum qubit state is highly volatile. Due to this volatility, the controller is also responsible for quantum error correction [188]. A controller consists of control and readout circuitry, service blocks, such as voltage, current, and frequency references [189], and a digital controller [190], as shown in Fig. 5.5. The service blocks consist of phase locked loop (PLL) oscillators, low noise amplifiers (LNA), analog-to-digital converter (ADC), digital-to-analog converter (DAC), and other circuitry.

The primary cryogenic technologies for the control circuitry of a quantum computer include semiconductor-based cryogenic CMOS, superconductive rapid single flux quantum (RSFQ) [33, 48, 191, 192], and adiabatic quantum flux parametron (AQFP) [33]. Circuits based on cryogenic CMOS technology dissipate less power as compared to room temperature CMOS, while delivering faster performance. CMOS has been reported to operate at temperatures ranging from 100 mK to room temperature [166]. At cryogenic temperatures, a MOSFET exhibits enhanced physical properties such as higher transient currents, negligible leakage currents, and increased subthreshold slope [34].

Niobium-based superconductive electronic systems require operation at a temperature below approximately 5 K (and frequently cooled to 4.2 K, helium boiling temperature) [193]. RSFQ and AQFP, two prominent superconductive technologies, have been steadily maturing over recent years [33, 194]. RSFQ systems exhibit 100

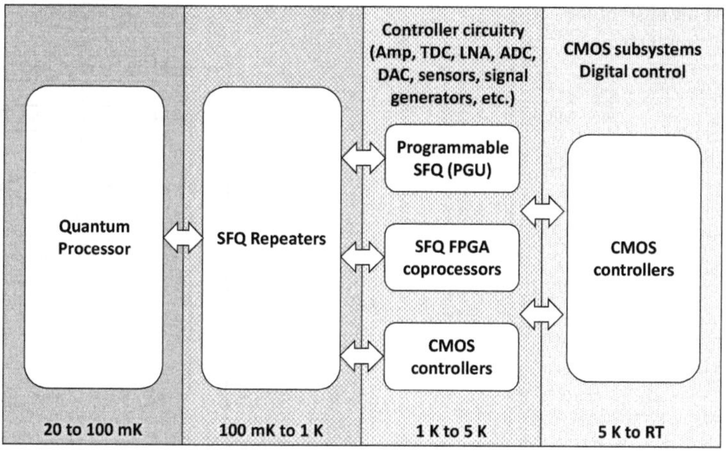

Fig. 5.5 Block diagram of hybrid quantum-classical computing system. The system consists of a quantum processor, SFQ repeaters, and controllers, each located at different temperatures. The controllers can be CMOS or SFQ

to 1,000 times less energy per computation as compared to CMOS, while operating at frequencies of hundreds of gigahertz. Systems based on AQFP technology dissipate even less power per computation, several orders of magnitude less than RSFQ, albeit at slower speeds [195]. Current manufacturing technology accommodates over 6,000 Josephson junctions (JJ) per mm^2 [196]. With the development of electronic design automation tools for superconductive electronics [197, 198], SFQ-based VLSI complexity systems are currently under development [33]. An 8 bit superconductive microprocessor operating at a frequency of 80 gigahertz has successfully been fabricated [199].

Despite these promising features, each of these technologies faces a significant challenge in the development of large scale systems. For example, SFQ memory is challenging due to the poor scaling of the inductors required in SFQ logic and memory. Cryogenic CMOS circuits, in turn, generate significant heat during operation, reducing the cooling efficiency. These issues can be overcome by combining multiple technologies within a single integrated system. The result of computing operations performed using an RSFQ controller can, for example, be stored in a CMOS static or dynamic random access memory (RAM) [34] or a magnetic tunnel junction (MTJ)-based magnetic RAM [200].

5.2 Optimization Setup

In the first step of the methodology, a graph of the system is generated, as described in Section 5.1.1. Any path connecting the initial stage of process S_1 to the final stage S_n determines the power consumption and delay of the system. The optimization

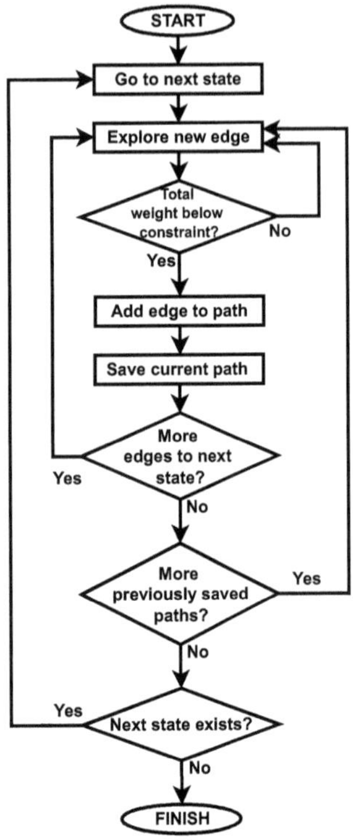

Fig. 5.6 Flowchart of the proposed algorithm

problem is to determine the most power efficient temperature set, while ensuring the delay of the system is below constraint D_{max}. A flowchart of the algorithm to determine all of the paths within the graph satisfying the delay constraint is illustrated in Fig. 5.6. The algorithm requires a matrix of delays D as an input, where entry $D_{i,j}$ denotes the delay of unit i at temperature T_j,

$$D = \begin{array}{c} \\ T_1 \\ T_j \\ T_m \end{array} \begin{array}{c} U_1 \quad U_k \quad U_n \\ \begin{bmatrix} D_{1,1} & \cdots & D_{1,n} \\ \vdots & \ddots & \vdots \\ D_{m,1} & \cdots & D_{m,n} \end{bmatrix} \end{array}. \tag{5.10}$$

The proposed algorithm is based on breadth-first search traversal of the process graph, starting from the source node. During the traversal, delay D of the partial path is compared to delay constraint D_{max}. If delay D is greater than D_{max}, the algorithm explores the next edge. Partial paths satisfying the delay constraint are recorded and the traversal continues. Upon completing the traversal process, a new path to the current node is treated as an input, and all of the edges are once again explored. After all of the paths from the source to sink are evaluated and the unwanted paths are discarded, the algorithm proceeds to the next node.

The algorithm includes a power weight attached to the edges. By determining the paths satisfying both the power P_{max} and delay D_{max} constraints, the memory usage and computational runtime are significantly reduced. The savings in memory and computational time are more significant if the possible path branch is removed earlier in the process. For example, for a process with ten units and ten possible temperatures, 10^{10} possible paths exist before the constraint check is performed. If all of the paths which start with $\pi = (U_1(T_1), \ldots)$ are removed in the first loop of the algorithm, 10^9 possible paths are removed, reducing the traversal time. The result of the algorithm is all possible paths from source to sink satisfy the delay constraint. The power flow between each unit is evaluated in the next step of the algorithm to determine the total power consumption of the path, as described in Section 5.1.2. Finally, the optimal temperature set is the set of temperatures consuming the least power.

5.3 Quantum Computing Case Study

A quantum computer is a combination of a quantum processor and an electronic controller [65]. A quantum processor uses quantum bits (qubits) to perform operations. Qubits operate at extremely low temperatures; typically, a few millikelvins [201]. An electronic controller reads out the signal and controls the quantum processor [65]. Existing quantum computers utilize classical electronic controllers operating at room temperature (RT) [176]. This approach, however, is challenging and expensive, as the number of qubits is expected to reach thousands and millions [176]. Establishing individual connections between millions of qubits and the controller circuitry operating at room temperature is infeasible due to the read complexity, cost, and signal performance of the interconnect [65, 176, 202]. It has therefore been suggested to utilize a classical CMOS electronic controller operating at cryogenic temperatures [65] or a SFQ controller operating below 4 K [33, 203, 204], which can be placed closer to the quantum processor. A block diagram of a hybrid quantum-classical system is depicted in Fig. 5.5.

While it is possible to operate most of the controller circuitry at temperatures below 4 K, the cooling capacity at these temperatures is often insufficient to efficiently dissipate the heat generated by the controller [65]. Partitioning the controller into higher and lower temperature domains may be more efficient. The proposed algorithm is used to determine the set of optimal temperatures for a hybrid superconductive quantum-classical computing system, as adapted from [203]. The system consists of different readout sensors, multiplexers, demultiplexers, and qubit drivers, placed at temperatures ranging from 20 mK to 4 K [203, 204]. A flowchart of the process is shown in Fig. 5.7. While the quantum computing system may consist of any number of units [203], the quantum computing system considered here consists of eleven units: two CMOS FPGAs, one for readout and one for control, four SFQ FPGAs, two SFQ pulse generating unit (PGU) systems, two SFQ quantum-classical interface (QCI) integrated circuits, and a quantum processor.

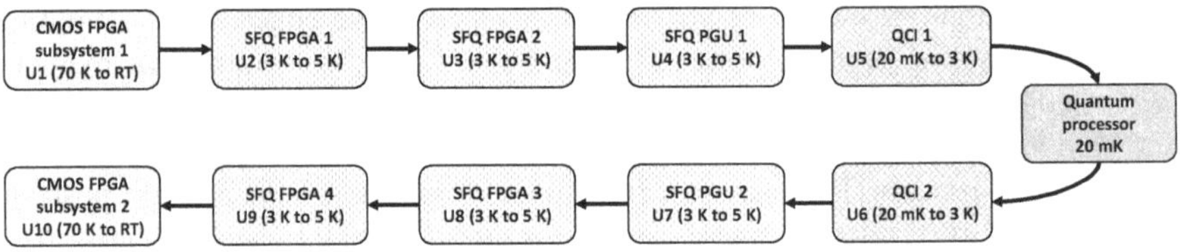

Fig. 5.7 Flowchart of temperature ranges for each function in a hybrid quantum-classical computing system

Table 5.2 Delay D_i and power P_i of each unit U_i in a hybrid quantum computer. The delay and power values are at the highest possible operating temperature of each unit

Unit	Delay	Power	Unit	Delay	Power
U_i	D, [fs]	P, [W]	U_i	D, [fs]	P, [W]
U_1	1,000	5	U_6	6	70
U_2	150	15	U_7	25	40
U_3	100	25	U_8	110	20
U_4	20	50	U_9	130	20
U_5	5	80	U_{10}	800	7

Each unit of a quantum computing system is placed within a different temperature domain. The superconductive qubits (quantum processor) are located at 20 mK. The SFQ co-processors for qubit control, error tracking, error correction, readout processing, and execution of the classical portion of the quantum algorithms are placed at temperatures ranging up to 5 K. The digital CMOS FPGA controllers are placed at a temperature ranging from 70 K to RT [203].

The delay and power consumption for each unit at different temperatures are required. These values are generated for this case study. Ten different operating temperatures are available for each unit, yielding 10^{10} possible paths in the process graph. The set of available temperatures is generated by linearly spacing the temperature ranges for each unit within the system. The CMOS circuits operate at a temperature ranging from 70 K to RT, SFQ circuits operate ranging from 3 K to 5 K, and SFQ QCI circuits operate over a range from 20 mK to 3 K [203]. A delay and power at each temperature are assigned to each unit. These numbers are assumed to be the mean value of the delay and power of each unit during operation and randomly generated assuming an exponential distribution over different temperatures. The delay and power for each unit are listed in Table 5.2. The total power consumption includes the power consumption of the refrigerators.

Any thermal interactions between the units are set by the interconnects between the units and the proximity of the units to each other. The interconnects between the SFQ integrated circuits and the QCI and between the QCI and the SFQ coprocessor are established via superconductive low heat loads and low crosstalk superconductive ribbon cables [203]. These connections maintain accurate timing and reliable transmission of the SFQ pulses. These connections and the nonideality of the refrigerators produce a thermal conductance between these units. A simplified thermal-electrical circuit model of the system is illustrated in Fig. 5.8. Ten different thermal resistances between the units are assumed. Each unit type is located

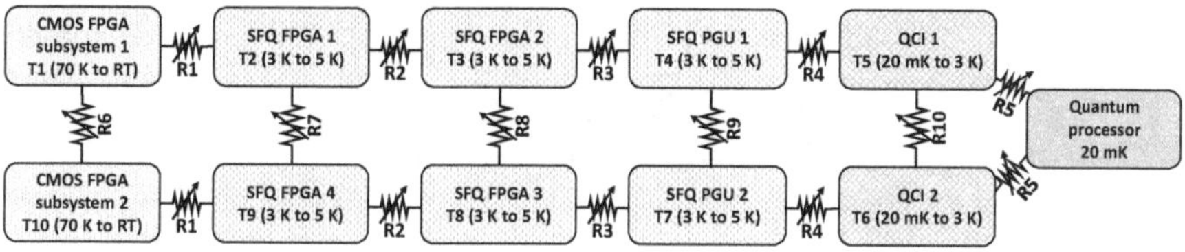

Fig. 5.8 Simplified thermal-electrical circuit model of the hybrid quantum computer. The thermal relationship between the units is described by the rheostats, which represent the dependence of the thermal resistance on temperature

Table 5.3 Thermal resistance of the hybrid quantum computer

Resistance	Ω_T [K/W]	Resistance	Ω_T [K/W]
R_1	60	R_6	30
R_2	150	R_7	50
R_3	200	R_8	100
R_4	400	R_9	150
R_5	600	R_{10}	300

at different temperatures. Thermal resistances R_1 to R_5 represent the thermal interaction between components, while R_6 to R_{10} denote the thermal conduction within a temperature domain. Due to the nature of the thermal resistance, which varies with temperature, the thermal resistance is adjusted based on the temperature of the connected components. Since the components primarily operate at cryogenic temperatures, the thermal resistance between components is assumed to linearly decrease with increasing temperature [205]. The value of the thermal resistances at 4 K is listed in Table 5.3.

A set of optimal temperatures is determined using the algorithm described in Section 5.2. The set of temperatures minimizing the total power while satisfying the delay constraint of 0.24 ps is determined. The algorithm is implemented in Python and executed on an Intel Core i7-9750H workstation with 8 GB RAM. The algorithm completes in 499.65 s for this case study. Sets of optimal temperatures, excluding the quantum processor, are listed in Table 5.4, where the most optimal set is highlighted in bold. The temperature of the CMOS FPGA and SFQ FPGA modules is the same in the top five most optimal temperature sets. The difference in performance is due to the difference in the temperature of the SFQ PGU and QCI modules. The power consumption of the optimal path with a delay constraint of 0.24 ps is 2,456 watts. Since the total consumption of the cooled components is 95.5 watts, most of the power is consumed by the refrigerators operating at cryogenic temperatures.

Table 5.4 Set of temperatures for a hybrid quantum computing system composed of two CMOS FPGAs, four SFQ FPGAcs, two SFQ PGUs, two SFQ QCIs, and a quantum processor. The most optimal (lowest power) set is highlighted in bold

Unit temperature, K										**Delay**	**Power**
U_1 CMOS1	U_2 FPGA1	U_3 FPGA2	U_4 PGU1	U_5 QCI1	U_6 QCI2	U_7 PGU2	U_8 FPGA3	U_9 FPGA4	U_{10} CMOS2	D_i, [fs]	P_i, [W]
70	**3**	**3**	**3.2**	**1.1**	**0.82**	**3**	**3**	**3**	**70**	**240.0**	**2,456**
70	3	3	3.2	0.82	0.82	3	3	3	70	239.2	2,775
70	3	3	3	0.82	0.82	3.2	3	3	70	239.7	3,475
70	3	3	3.2	1.4	0.51	3	3	3	70	239.8	3,698
70	3	3	3.2	1.1	0.51	3	3	3	70	239.2	3,857

5.4 Conclusions

Hybrid cryogenic computing systems are an emerging technology motivated primarily by cryogenic high performance computing and quantum computing networks. The operating temperature of the circuit components affects the performance, cooling power, and dissipated power. Selecting the appropriate operating temperature is therefore crucial to minimizing the total power dissipated by the system while maintaining correct functionality and performance.

A methodology for thermal optimization of cryogenic computing systems with multiple temperature zones is presented in this chapter. The methodology is validated on a practical case study where the individual temperature of an eleven unit system is optimized. The power consumption of a quantum computing system is minimized while satisfying the target delay constraint. A multigraph representation describes the relationship among the temperature, delay, and power of a system. All possible conditions of the system are represented by this multigraph. The total cooling power is described by a thermal model of the system, which includes a variable thermal conductance between each unit within the system. The proposed algorithm is applied to the case study, and the temperature of each component that minimizes the total system power dissipation is determined while satisfying target performance constraints.

Chapter 6
Heat Load Efficiency in Multi-Temperature Zones

The escalating demand for high performance computing (HPC) in recent years, propelled by the rise of computationally demanding applications such as cloud computing, has introduced several challenges. These challenges include energy efficiency, thermal management, and system performance. Data centers, which form the backbone of HPC systems, consume significant energy, ranging from tens to hundreds of megawatts [169]. The global energy consumption for HPC is estimated at approximately 200 TWh annually and is expected to quadruple by 2030 [49].

To sustain this rapid expansion, alternative computational technologies are being explored. Cryogenic technologies have emerged as a promising solution, capable of substantially reducing power consumption in large scale computing systems, including the energy associated with refrigeration [36, 45]. The cooling capacity of these cryogenic technologies at 4 K may however be insufficient for efficient heat dissipation [65]. Furthermore, the available cooling power varies at different temperatures. As shown in Chapter 5, the available cooling power per kilowatt input power differs across temperatures. Consequently, certain circuits may benefit from operating at lower temperatures, whereas other circuitry could achieve satisfactory performance at higher temperatures.

Optimizing the temperature of each subsystem within a computing system can enhance overall performance, power efficiency, and heat management. For instance, changing the temperature of a subsystem within a cryogenic system permits the integration of different technologies and functionalities at distinct temperature zones within a cryocooler. This strategic placement can lead to a reduction in refrigeration costs. Raising the temperature of a subsystem may however lead to increased latency and power dissipation.

Furthermore, the refrigeration characteristics of adjacent subsystems, particularly those operating at lower temperatures, can be affected if these systems are not adequately thermally isolated. The heat emanating from a subsystem operating at a higher temperature could inadvertently affect the cooling efficiency of neighboring lower temperature systems, particularly if these systems are interconnected.

N. Zhuldassov, E. G. Friedman, *Cryogenic Microelectronic Systems for Ultra-Low Energy and Enhanced Performance*, https://doi.org/10.1007/978-3-031-90754-8_6

Multiple research efforts have focused on placing different technologies and functions across different temperature stages within a refrigeration system [174, 175]. For instance, a hybrid temperature system integrated in the Sumitomo SRDK-101DP-11C cryocooler is described in [174]. This system features a 4 K stage for low temperature superconductive circuits and a 60 K stage for higher temperature semiconductor circuits, such as analog filters and low noise amplifiers, while other electronics are maintained at room temperature (RT). These studies, however, often overlook the full spectrum of temperatures accessible within a given stage. In this system, for example, although the second stage is set to 60 K, the actual temperature range can accommodate temperatures from 60 K to 80 K. Such insights highlight the importance of considering the entire range of feasible temperatures for each stage to optimize the placement and operation of different components within a cryogenic system.

Exploiting the full range of temperatures available in each stage of a cryocooler can enhance the performance of a computing system. A new approach is proposed to optimize the operating temperature of each component within a cryogenic system by minimizing the total power consumption of the system while ensuring that the performance remains within acceptable parameters. The particular thermal characteristics of each stage within a cryocooler is considered in the optimization process, allowing for a more efficient and effective distribution of components based on thermal and power requirements.

In Chapter 5, the temperature variability within each stage is exploited; the dependence of the power consumption of each cooler stage on the temperature of a neighboring stage is, however, not considered. A computing unit can be placed inside a refrigerator equipped with multiple cooling stages; for example, three stages, where the temperature of stage 1 is 150 K, stage 2 is 70 K, and stage 3 is 4 K. In contrast, if the same computer is placed inside a refrigerator with 120 K at stage 1, 60 K at stage 2, and 4 K at stage 3, the performance and power consumption will be different. The advancement described here considers, for example, the variation in power consumption of the third stage of a cryocooler that is dependent on the temperature of the first and second stages. Thus, this methodology improves the operation of the overall system rather than individually optimizing the operation within each temperature zone.

In addition, the application of cubic spline interpolation is introduced to estimate the power consumption and performance at different temperatures, and a method to speed up the algorithm is also presented. The proposed methodology is validated on a case study of a superconductive cloud computing system. The cloud system utilizes both CMOS and superconductive logic for computation and storage.

The chapter is organized as follows: in Section 6.1, insight into the organization of cryogenic coolers are described, the problem is formulated, and the thermal behavior of the system is discussed. The proposed methodology is described in Section 6.2. An example case study, a hybrid superconductive/semiconductor cloud computing system, optimized using the proposed methodology, is presented in Section 6.3. Some conclusions are offered in Section 6.4.

6.1 Background of Cryogenic Cooling and Thermal Optimization

The effective integration of cryogenic computing systems necessitates electronic circuits operating at different cryogenic temperatures [65]. The primary goal of the proposed methodology is to identify a set of optimal temperatures for each component, minimizing the total power consumption or delay of a cryogenic system.

A methodology is proposed to determine the optimal temperature of the different components of an electronic system, as shown in Fig. 6.1. A graph of the system is initially constructed using measured or interpolated power and delay values of each component at different temperatures. Cubic spline interpolation [206] is employed for approximating and interpolating the variable weights within the graph. An algorithm based on graph theory [178] is utilized to determine the set of optimal temperatures. In this algorithm, the power and delay weights in the graph are adjusted after each state based on the temperature from the preceding state, as described in Section 6.2. To determine the optimal temperature within each zone, the algorithm is executed twice, with the second iteration refining the available temperature ranges based on the results of the first pass. Once the appropriate temperature set satisfying the constraints is identified, a thermal model of the system is developed to assess the heat flow (or power transfer) between units. This heat flow is influenced by the thermal conductance between the units and

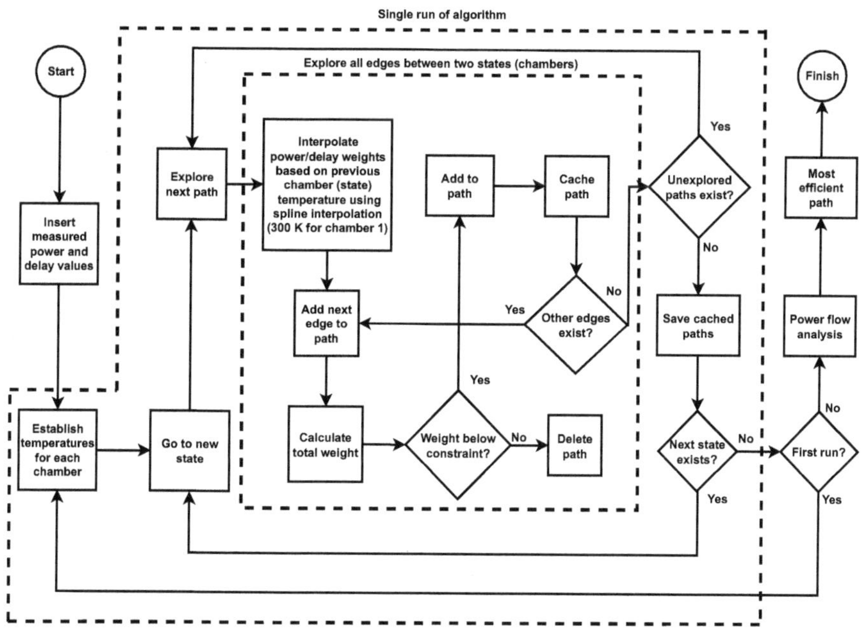

Fig. 6.1 Flowchart of the proposed methodology

the temperature of the interconnecting cables. Furthermore, the leakage power is estimated based on the heat flow; particularly, the power loss due to additional cooling required in lower temperature components resulting from heat transfer from higher temperature components. The leakage is contingent upon the thermal conductivity of the cables connecting the components. Inclusion of this effect significantly expands the generality of the work described in Chapter 5. The overall power consumption for a given set of temperatures thus includes the leakage power between different temperature zones. To achieve optimal system operation, considering delay, power, and the difference in temperature induced heat flow among components, the temperature of each component is determined.

The rest of this section is organized as follows. Background information describing cryogenic coolers is discussed in Section 6.1.1. The problem formulation based on a graph theoretic approach is described in Section 6.1.2. Cubic spline interpolation is summarized in Section 6.1.3. A thermal model of the system is discussed in Section 6.1.4.

6.1.1 Cryogenic Coolers

Cryogenic coolers are essential in low temperature applications such as superconductive devices [33, 48, 194, 198], quantum computing, medical systems, and materials research. Several types of cryocoolers exist, such as Gifford-McMahon cryocoolers, pulse-tube cryocoolers, and dilution refrigerators [69]. Dilution refrigerators can achieve tens to hundreds of millikelvins. Achieving these extremely low temperatures requires multiple stages of cooling, each stage being progressively colder than the previous stage. The overall structure of a cryogenic cooler typically consists of multiple temperature chambers, each containing a refrigerant that cools the following stage. For example, the PT415 pulse tube cryocooler from Cryomech [207] has a 50 K stage with a temperature range between 35 K to 77 K, and a 4 K stage with a temperature range from 3 K to 18 K.

The first stage of a cryogenic cooler typically contains both air cooled and water cooled compressors, with temperatures ranging between approximately 120 K to room temperature [208, 209]. The next stage usually operates within a temperature range between 50 K to 120 K, and is cooled by liquid nitrogen [207–209]. To reach lower temperatures, neon can be used in the stage between 20 K and 50 K [69]. Liquid helium is used as a refrigerant in the next stage to achieve a temperature between 3 K and 5 K [69, 170, 208, 209]. To reach the lowest temperatures, down to tens to hundreds of millikelvins, $^{3}He/^{4}He$ dilution refrigerators are used, where Helium-3 and Helium-4 isotopes are mixed [69, 170]. Some examples of commonly used coolants and related operating temperatures are noted in Fig. 6.2 [69–72].

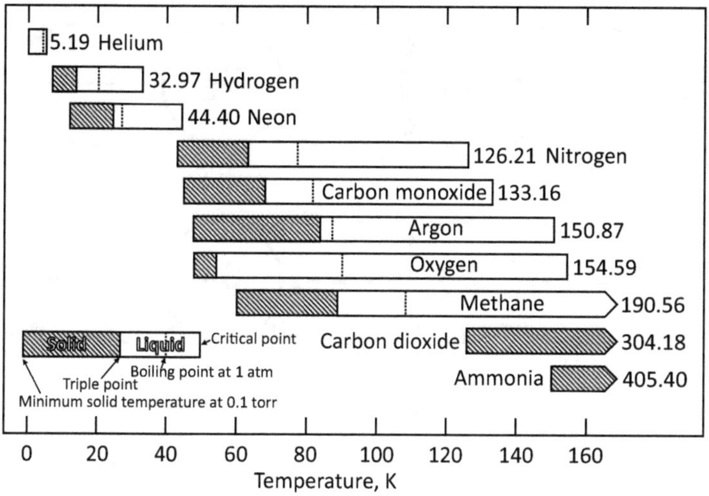

Fig. 6.2 Commonly used coolants and relevant operating temperatures [69–72]

6.1.2 Formulation of Thermal Optimization Problem

The objective is to choose the ideal operating temperature at each stage of the process. The optimization process for choosing the zone specific temperatures can be conceptualized as a directed acyclic multiweighted multigraph, with one variable and one constant weight at each edge $G := \langle S, C, W \rangle$. Here, $S = S_1, S_2, \ldots, S_n$ is a finite set of states defining a specific instance of the temperature optimization problem. The edges, denoted by C, represent chambers within a refrigeration unit, each containing one or more computing units. For each chamber i, the subset $C_i \subseteq C$ corresponds to parallel edges. Typical refrigeration systems operate at specific temperatures, such as liquid helium temperature (LHT) or liquid nitrogen temperature (LNT), represented by the set of available temperatures $T = T_1, T_2, \ldots, T_j$. Each chamber operating at a different temperature during each step is denoted by $c_{i,j} \in C_i$. $W_{i,j} := \langle p(T_{i-1,j}), d \rangle \in \mathbb{R}^2_{>0}$, where $p(T_{i-1,j})$ and d represent, respectively, the power consumption and delay of a unit at a specific temperature. The power consumption is a variable weight dependent on the previous edge weight along the chosen path.

A path π in this process graph, representing a set of operating temperatures for each unit, links the source to the sink. It is represented as

$$\pi = \big(C_1(T_j), C_2(T_j), \ldots, C_i(T_j)\big) . \tag{6.1}$$

The power consumption of a process is the sum of the power weights along a path, $P(\pi) = p_1 + p_2 + \cdots + p_{n-1}$, with each weight representing the power consumption of a unit. Similarly, the total delay of the process, $D(\pi) = d_1 + d_2 + \cdots + d_{n-1}$, is the sum of the delay weights of the units along the path. The temperature optimization

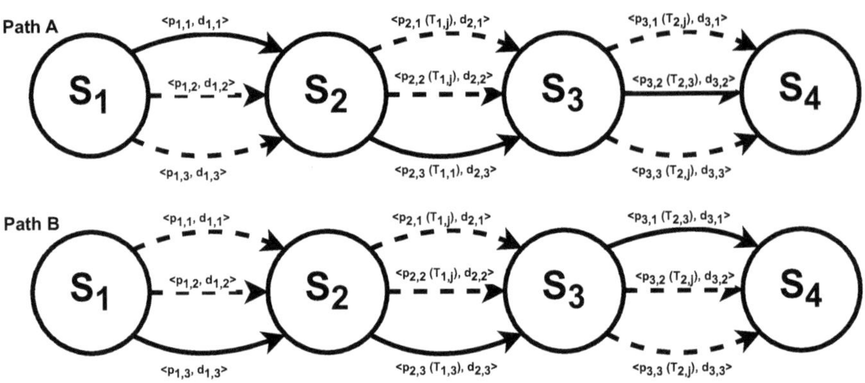

Fig. 6.3 Example of the temperature design process using a multiweighted multigraph with variable edges. The edges between two states denote a chamber within a refrigerator at different temperatures. Two paths, A and B, are shown highlighted in bold. The power consumption of path A is $P(\pi_A) = p_{1,1} + p_{2,3}(T_{1,1}) + p_{3,2}(T_{2,3})$, and the power consumption of path B is $P(\pi_B) = p_{1,3} + p_{2,3}(T_{1,3}) + p_{3,1}(T_{2,3})$. Note that although the edge between states S_2 and S_3 is the same in both paths, the power weight of the edge is different. The power consumption is determined by the temperature of the previous chamber, which is denoted by $T_{1,1}$ in the edge between states S_2 and S_3 in path A, and by $T_{1,3}$ in path B. The delay of path A and path B is, respectively, $D(\pi_A) = d_{1,1} + d_{2,3} + d_{3,2}$ and $D(\pi_B) = d_{1,3} + d_{2,3} + d_{3,1}$

problem is to find a path that minimizes the total power $P(\pi)$ while keeping the total delay $D(\pi)$ below a maximum limit D_{max}.

$$\text{Minimize: } P(\pi), \tag{6.2}$$

$$\text{subject to: } D(\pi) \leq D_{max}. \tag{6.3}$$

An example with three units and four states is illustrated in Fig. 6.3. Each unit can operate at three distinct temperatures, indicated by the parallel edges between adjacent states. Paths A and B, highlighted in bold, represent two potential configurations. Path A, $\pi_A = (C_{1,1}, C_{2,3}, C_{3,2})$, and Path B, $\pi_B = (C_{1,3}, C_{2,3}, C_{3,1})$, correspond to different temperature settings for the chambers. The power consumption and delay of these paths are based on the respective weight of the edges along the path.

6.1.3 *Cubic Spline Interpolation*

The cubic spline interpolation technique is used here [206] to interpolate the power consumption of a unit within a cooler chamber. An interpolating function is used, as it is intractable to manually define a graph with fixed weights for all possible combinations of temperatures. Cubic spline interpolation is particularly useful in this application, as the cubic curvature and slope of the resulting interpolation is similar to the exponential relationship between temperature and power consumption

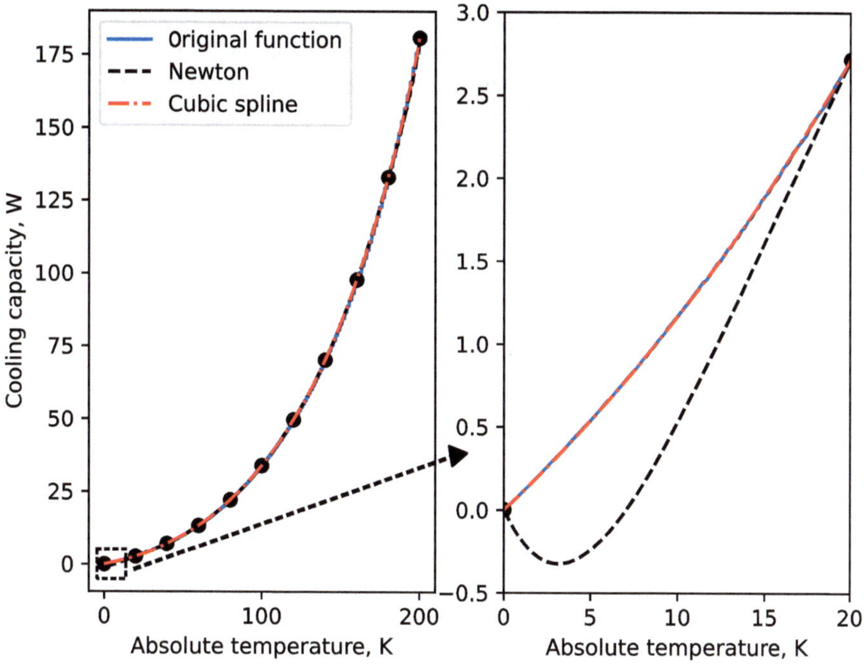

(a) (b)

Fig. 6.4 Techniques to interpolate the available cooling power per kW input power, as depicted in Fig. 5.1 for (a) the 20 mK to 200 K range and, (b) the 20 mK to 20 K range. Cubic spline interpolation closely resembles the original curve as compared to Newton polynomial interpolation

[36]. Other linear interpolation techniques, such as Lagrange polynomial and Newton polynomial, are less suitable due to higher errors, as shown in Fig. 6.4.

For a given set of $n + 1$ data points (x_i, y_i) with no repeating x_i and $a = x_0 < x_1 < \cdots < x_n = b$, spline $S(x)$ is a function satisfying:

1. $S(x) \in C^2[a, b]$;
2. On each $[x_{i-1}, x_i]$, $S(x)$ is polynomial of degree 3, where $i = 1, \ldots, n$;
3. $S(x) = y_i, \forall i = 0, 1, \ldots, n$.

Interpolated function $S(x)$ becomes

$$S(x) = \begin{cases} C_1(x), & x_0 \leq x \leq x_1 \\ \cdots \\ C_i(x), & x_{i-1} \leq x \leq x_i \\ \cdots \\ C_n(x), & x_{n-1} \leq x \leq x_n, \end{cases} \tag{6.4}$$

where each $C_i = a_i + b_i x + c_i x^2 + d_i x^3$ is a cubic function with $d_i \neq 0$ for $i = 1, \dots, n$. An example of cubic spline interpolation and Newton polynomial interpolation is shown in Fig. 6.4. Interpolation techniques are used to approximate the available cooling power per one kilowatt of input power [35]. Note that in this case cubic spline interpolation exhibits less error than Newton polynomial interpolation.

6.1.4 Thermal Model

In addition to the heat generated by the units, the power consumed along a path also includes leakage power resulting from temperature differences, transferred through the connecting cables [35]. As the thermal resistance of cable materials changes with absolute temperature, adjustments are made for more precise heat flow characterization between units [183, 184]. The thermal resistance changes based on the material type, quality, and purity, and can be either linear, exponential, or logarithmic [183, 184, 210].

For cryogenic applications, specialized cables such as CryoCoax BCB016, BCB019, and BCB029, made of stainless steel and beryllium copper, are employed [211]. These materials demonstrate increasing thermal conductivity with rising temperature [185–187]. While the thermal conductivity of beryllium copper can be linearly approximated [187], the conductivity of stainless steel is best represented through a dual-line approximation [186], as depicted in Fig. 6.5.

The thermal conductivity of different cables is considered to construct thermal circuits, analogous to electrical circuits, as described in Chapter 5. The heat flow within the system, denoted as q_T, is represented by a set of linear equations,

$$q_T = \begin{array}{c} \\ U_1 \\ U_k \\ U_n \end{array} \begin{array}{c} \begin{array}{ccc} U_1 & U_k & U_n \end{array} \\ \begin{bmatrix} \frac{\Delta T_{1,1}}{R_{1,1}} & \cdots & \frac{\Delta T_{1,n}}{R_{1,n}} \\ \vdots & \ddots & \vdots \\ \frac{\Delta T_{n,1}}{R_{n,1}} & \cdots & \frac{\Delta T_{n,n}}{R_{n,n}} \end{bmatrix} \end{array} . \tag{6.5}$$

Here, heat flow is determined by the temperature difference ΔT and the thermal resistance R between the units. The power transfer to or from each unit is determined by summing the heat flow along each row of this matrix,

$$\Delta P = q_T \, \mathbf{1}_n, \tag{6.6}$$

where $\mathbf{1}_n$ is a column vector of ones,

$$\mathbf{1}_n = [1, \dots, 1]^{\mathsf{T}} . \tag{6.7}$$

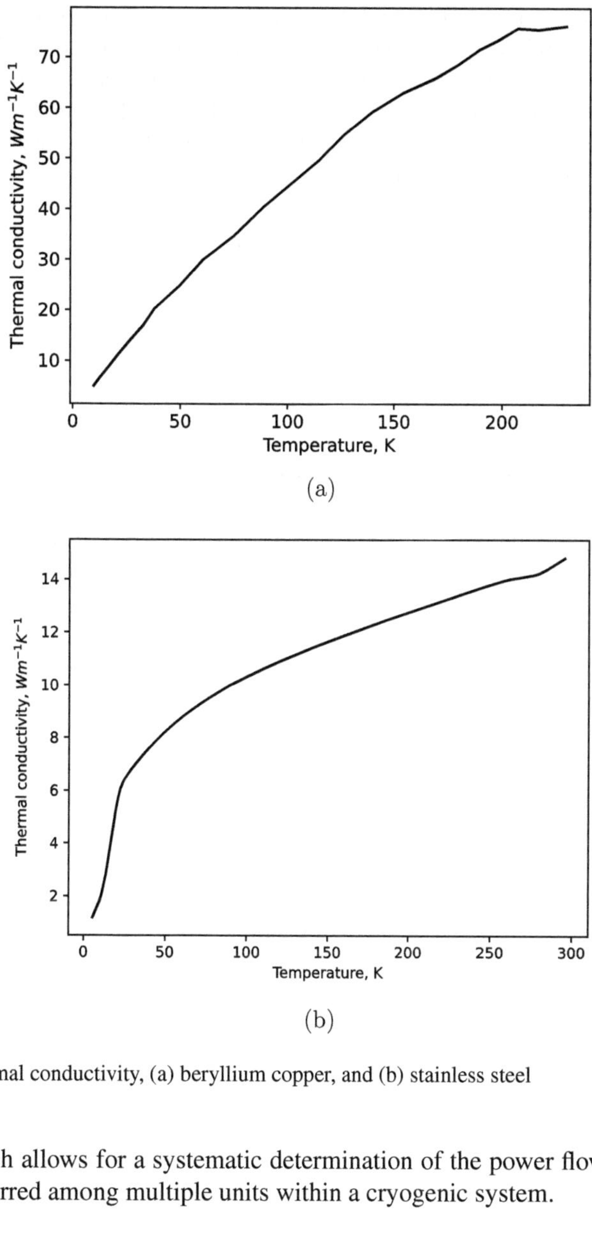

Fig. 6.5 Thermal conductivity, (a) beryllium copper, and (b) stainless steel

This approach allows for a systematic determination of the power flow due to heat being transferred among multiple units within a cryogenic system.

6.2 Optimization Setup

The initial phase of this methodology requires a system graph, as described in Section 6.1.2. In this graph, each path from the starting state S_1 to the final state S_n

affects the power consumption and delay. The optimization problem is to identify the most energy efficient set of temperatures, ensuring that the overall delay of the system remains below D_{max}. A flowchart of the algorithm required to identify all of the graph paths satisfying the delay constraint is illustrated in Fig. 6.1.

The algorithm is initiated with an input delay matrix D, where each element $D_{i,j}$ denotes the delay of unit i at temperature T_j,

$$D = \begin{array}{c} \\ T_1 \\ T_j \\ T_m \end{array} \begin{array}{c} \begin{array}{ccc} C_1 & C_k & C_n \end{array} \\ \begin{bmatrix} D_{1,1} & \cdots & D_{1,n} \\ \vdots & \ddots & \vdots \\ D_{m,1} & \cdots & D_{m,n} \end{bmatrix} \end{array}. \tag{6.8}$$

Delay values within a temperature range are interpolated via cubic spline interpolation, as outlined in Section 6.1.3. The power weights are similarly estimated.

Breadth-first search is employed in the algorithm to traverse the process graph from the source node, comparing the delay of each partial path to D_{max}. Paths satisfying the delay constraint are tracked, and upon exploring all of the edges from a node, the algorithm proceeds to the next node. Since the power weights are variable, the weight is reliant on the power weight of the edges preceding the current state. The initial and interpolated power weights are adjusted to obtain new power weights. Newton's law of cooling is applied, based on the previous edge, to achieve the required adjustments.

Memory usage and computational time are optimized by a two step approach. First, those paths not satisfying both power and delay constraints are discarded. Second, running the algorithm twice refines the temperature range in subsequent steps, reducing the complexity from $O(n^4)$ to $O(n^2)$. This technique maintains precision while enhancing the computational efficiency.

The algorithm is not limited by a constant number of refrigerator chambers, and can determine the most optimal number of temperature zones. For example, if certain circuits are placed within different temperature zones, a state for each temperature zone is added to the graph. Rather than individually managing each temperature zone, a combined temperature range that encompasses all of the zones for these new states is used.

Once all of the potential paths satisfying the delay limit are identified, the next step evaluates the power flow between units to determine the total power consumption along each path. This process is described in Section 6.1.4. The set of optimal temperatures is selected based on the lowest power dissipation.

6.3 Case Study: Cryogenic Cloud Computing

Cloud computing provides computational, software, and storage services [212]. Cloud computing centers are stationary and mostly operate at room temperature. Cloud computing systems can be placed within a cryogenic environment to utilize cryoCMOS and superconductive logic to enhance computational speeds while reducing energy consumption and heat load [33–35, 191, 192]. A block diagram of a cloud computing system operating within different temperature zones is depicted in Fig. 6.6. Note that the system is partitioned into different temperature zones based on the cooling requirements.

In many instances, operating the majority of a system at temperatures below 4 K is feasible. However, at these lower temperatures, the cooling capacity often proves inadequate for effectively dissipating the heat produced by the CMOS components within the system [65]. Consequently, dividing a system into domains with higher and lower temperatures may result in more efficient power and delay characteristics. The proposed algorithm described herein is used to determine the set of optimal temperatures for a cryogenic cloud computing system. The system is comprised of different components such as multiplexers, demultiplexers, arithmetic logic units, registers, and other computing units, which are strategically placed at different temperatures from 3 K to 300 K. Depending upon the dissipated heat of the circuits, some of the components are placed in the fourth stage, with a temperature domain between 3 K to 5 K, cooled by liquid helium. Other CMOS circuitry is partitioned into three temperature domains ranging between 20 K to RT. The CMOS circuitry dissipating the least heat, such as the CMOS amplifiers, time-to-digital converters, low noise amplifiers, sensors, analog-to-digital converters, digital-to-analog converters, and signal generators, are placed within the third stage of the cryocooler, in the 20 K to 45 K temperature domain, cooled by liquid or solid

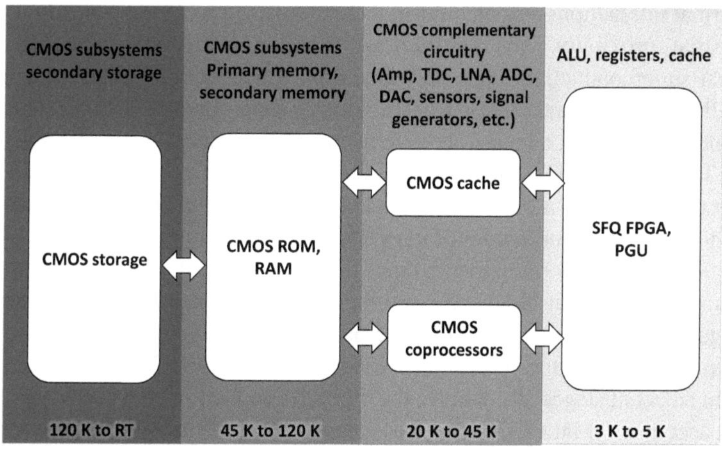

Fig. 6.6 Block diagram of cloud computing system placed at different temperature zones

Table 6.1 Delay D_i and power P_i of each circuit group within a cryogenic cloud computing system at the maximum feasible operating temperature for each unit

Refrigerator Stage	**Delay** D, [ns]	**Power** P, [kW]
Stage 1	1000	16
Stage 2	150	10
Stage 3	100	4
Stage 4	20	1.5

neon. The primary and secondary memory, such as RAM, are placed within the next refrigerator chamber, the second stage, cooled by liquid or solid nitrogen within the 45 K to 120 K temperature domain. The CMOS circuitry consuming the most heat, such as long term storage, is placed in the first stage of the cryocooler within the 120 K to RT domain chamber, which is cooled by an air and/or water cooled compressor [208, 209].

The optimization process utilizes the delay and power consumption measurements of each unit at different temperatures. Estimates are generated for this case study. Four different operating temperatures are available in each chamber. The set of available temperatures is generated by linearly spacing the temperature range for each device within each system. Each computing unit is assigned delay and power values at each temperature, which are considered to be the average value of the delay and power. In this case study, power consumption and delay measurements at ten distinct temperatures are used. These measurements are subsequently employed to estimate the power consumption and delay at different temperature intervals through cubic spline interpolation, as described in Section 6.1.3. The delay and power for each circuit group placed in separate refrigerator chambers are listed in Table 6.1. The overall power consumption encompasses not only the power dissipated by the computing units but also includes the energy consumed by the refrigeration systems. These power consumption and delay values are used within a graph of the system. The paths in this graph are explored, as described in Section 6.2 and Fig. 6.1.

Thermal interactions among units in the system are governed by the interconnects and spatial proximity. The connections between refrigerator chambers utilize low heat superconductive loads and superconductive ribbon cables with minimal crosstalk for reliable and precise transmission of superconductive logic signals [33]. Other necessary interconnects are established through cryocoax cryogenic cables [211]. These connections, combined with the nonideal cooling efficiency of the refrigerators, produce a thermal conductance between the chambers. A simplified thermal-electrical circuit model of the system is depicted in Fig. 6.7.

The system incorporates five distinct thermal resistances influencing the interactions between circuit blocks. Some blocks are situated within the same stage of a refrigerator. Thermal interactions are assumed to only occur between different stages of the refrigerator. Thermal resistances R_1, R_2, and R_3 signify interactions between adjacent stages, R_4 denotes the thermal conduction between stages 1 and 3, and R_5 accounts for interactions between stages 2 and 4. Since the thermal resistance changes with temperature, the resistance is adjusted based on the temperature of the interconnected components. Since the components operate at cryogenic

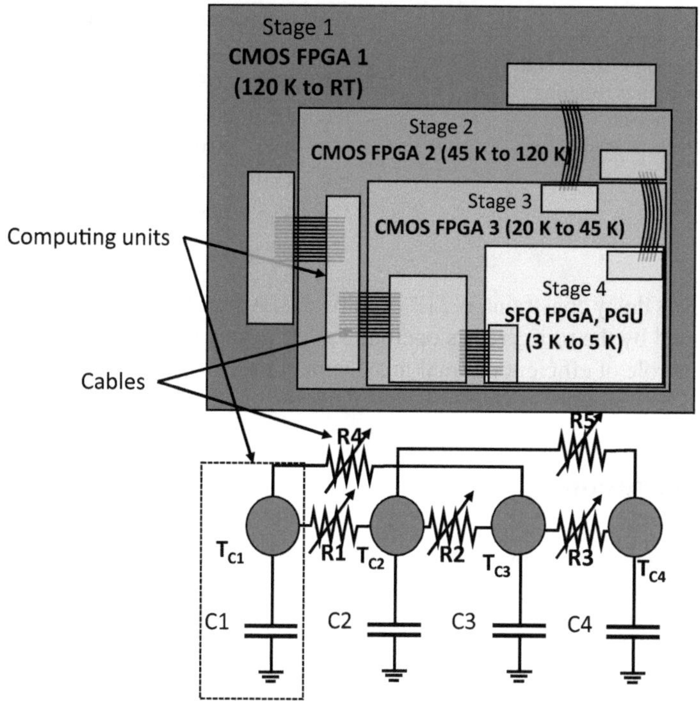

Fig. 6.7 Simplified thermal-electrical circuit model of a cryogenic cloud computing system. The thermal relationship between the chambers due to the temperature difference are represented by resistors, which characterize the dependence of the thermal resistance on temperature

Table 6.2 Thermal resistances of the cryogenic cloud computing system

Resistance	Ω_T [K/W]	Resistance	Ω_T [K/W]
R_1	60	R_4	310
R_2	150	R_5	450
R_3	200		

temperatures, the thermal resistance between these components linearly decreases as the temperature rises [205]. The thermal resistances at 4 K are listed in Table 6.2.

The proposed algorithm, described in Section 6.2, is employed to identify the set of temperatures that minimize total power consumption while adhering to a maximum delay constraint of 0.135 μs. The algorithm is implemented using Python on an Intel Core i7-9750H workstation equipped with 8 GB RAM. In this case study, the execution time is 0.65 seconds. The sets of optimal temperatures identified by the algorithm are listed in Table 6.3. The most power efficient set is highlighted in bold. Notably, in the first, second, and fourth most optimal temperature sets, the temperature of the CMOS modules in the first three stages remains constant. The variation in performance in these sets are attributed to the different temperatures of the nonCMOS modules. The power consumption of the optimal path, which satisfies

Table 6.3 Set of temperatures for a cryogenic cloud computing system. The set of temperatures resulting in the lowest power consumption is in bold

Stage temperature, K				Delay	Power
C_1	C_2	C_3	C_4	D_i, [ns]	P_i, [kW]
120	**45.0**	**20.0**	**4.38**	**134.83**	**218.71**
120	45.0	20.0	4.36	134.55	219.48
120	45.9	20.0	4.36	134.98	219.62
120	45.0	20.0	4.33	134.28	220.26
120	45.9	20.0	4.33	134.72	220.41

the 0.135 μs delay constraint, is 218.71 kilowatts. A significant portion of this power is consumed by the refrigerators operating at cryogenic temperatures, underscoring the critical role of efficient thermal management in such systems.

6.4 Conclusions

Cloud computing data centers are widely used. These cloud computing systems can exploit cryogenic operation to achieve high performance computing while dissipating significantly lower energy. The operating temperature profoundly influences the performance, cooling capacity, and power dissipation of the circuit components. It is therefore essential to carefully choose the operating temperatures to reduce the total power dissipation (and heat load) of the overall system while ensuring the proper function and enhancing the performance.

The methodology outlined in this chapter focuses on the thermal optimization of cryogenic computing systems operating across different temperature zones. Unlike previous approaches that focus on individually optimizing each subsystem, the proposed methodology considers the operation of the entire system including the mutual dependence between temperature zones. The total power dissipation of a system is minimized while maintaining the performance within predefined delay constraints. A practical case study is evaluated to validate the methodology in which the temperature of the individual subsystems within a cryogenic computing system, located in a four stage refrigerator, is determined. The temperature, delay, and power of the system are represented by a multigraph with variable power and constant delay weights. This multigraph contains all possible states of the system. The power consumption of an optimized cryogenic cloud computing system is 218.71 kilowatts and the total delay is 0.135 μs. Thermal leakage through the interconnecting cables between separate chambers can be a significant issue. An estimate of the total cooling power includes this leakage within the thermal model of the system.

Chapter 7
Integrated Multi-Temperature Cryogenic Computing Systems

The computing industry is advancing along several distinct trajectories, many of which operate at cryogenic temperatures. Two examples of these trajectories are superconductive digital computing systems and quantum computing systems [201, 213]. Superconductive digital electronics is a highly promising beyond-CMOS technology, offering both ultra-low power consumption and ultra-high speed [198, 213, 214]. These systems operate at cryogenic temperatures [192], as superconductivity is achieved in niobium at temperatures below 9.3 K and frequently cooled to liquid helium temperature (4.2 K).

Despite the low energy efficiency of operating at cryogenic temperatures in small scale applications, a significant increase in efficiency is observed as the scale of the system being cooled increases. For example, at 4 K, the energy efficiency can range from 1% of Carnot efficiency [43] for small units to as high as 35% for large scale liquefaction plants [44]. This enhancement in energy efficiency allows large scale cloud computing centers to operate at greater performance and higher energy efficiency [45].

These systems can be partitioned into multiple computing units, each of which can be placed within a different temperature zone. Quantum computers, for example, apart from the quantum processor operating at millikelvin temperatures, may consist of single flux quantum (SFQ) circuitry [213], CMOS circuitry, and various support blocks such as phase locked loop oscillators and low noise amplifiers [203]. Distributed computing systems are composed of thousands or even millions of servers distributed across multiple data centers.

Each unit exhibits different power consumption profiles, performance characteristics, and thermal load requirements. The available cooling power per kilowatt input differs across temperatures [35]. Consequently, certain circuits may benefit from operating at lower temperatures, whereas other circuitry would achieve satisfactory performance at higher temperatures. Increasing energy efficiency at cryogenic temperatures combined with the higher cooling efficiency of large scale refrigeration results in higher power efficiency and improved performance as compared to

N. Zhuldassov, E. G. Friedman, *Cryogenic Microelectronic Systems for Ultra-Low Energy and Enhanced Performance*, https://doi.org/10.1007/978-3-031-90754-8_7

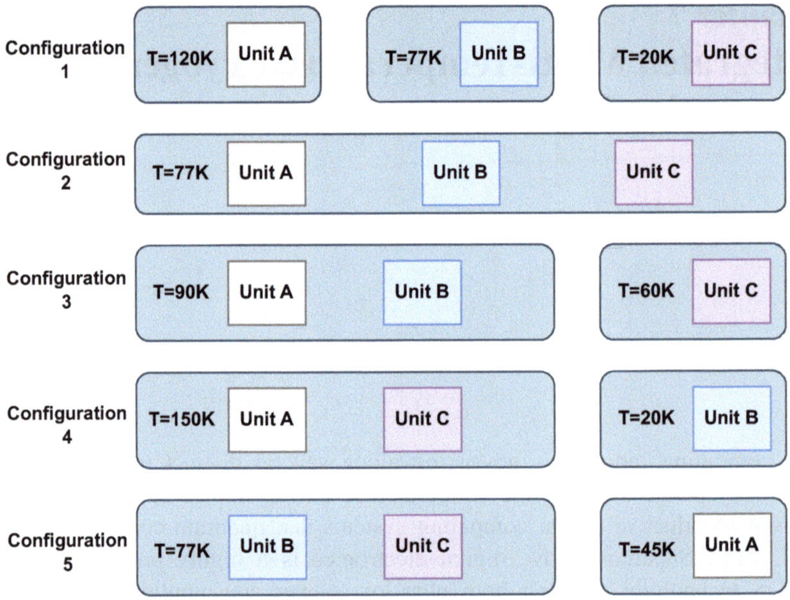

Fig. 7.1 Five different possible refrigerator configurations available for a three unit system

operating at room temperature [36, 45]. Previous approaches on managing this problem have relied on a pre-determined number of refrigerator stages and unit configurations [35, 37].

The methodology proposed here addresses the limitations of earlier approaches by optimizing the number of refrigeration stages based on the specific requirements of the system. Rather than relying on fixed unit assignments per refrigeration stage, the proposed method introduces flexibility by enabling units to be grouped based on the local performance and power profiles, ensuring that each group operates at the most energy efficient temperature. Both the number of stages and the configuration of the units within those stages are determined based on performance constraints, while maximizing energy efficiency and minimizing power consumption.

As an example, consider an idealized system composed of three units with five different configurations, as shown in Fig. 7.1. Each of these configurations can operate at a different set of temperatures, producing different performance and power consumption characteristics. The proposed methodology would determine the optimal configuration and set of temperatures.

The chapter is organized as follows: A brief review of previous work and an overview of certain applications are provided in Section 7.1. A formulation of the problem and proposed solution are discussed in Section 7.2. A discussion of the algorithm complexity is also presented, followed by suggestions on how to accelerate the algorithm. Two case studies, a hybrid superconductive/semiconductor distributed computing system and a quantum computing system, are optimized using the proposed methodology, and are discussed in Section 7.3. Some conclusions are offered in Section 7.4.

7.1 Background

This section is organized as follows: insight into previous work is provided in Section 7.1.1. Example applications for the proposed methodology are reviewed in Section 7.1.2.

7.1.1 Prior Work

Several earlier research efforts focused on allocating different technologies and functions across different temperature stages within refrigeration chambers [174, 175]. For example, a multi-temperature system integrated with the Sumitomo SRDK-101DP-11C cryocooler is described in [174]. In this system, low temperature superconductive circuits are placed within a 4 K chamber, higher temperature semiconductor circuits, such as analog filters and low noise amplifiers, are placed in a 60 K stage, while other electronics operate at room temperature (RT). These studies, however, often overlook the full spectrum of temperatures that each stage can provide. In these systems, although the second stage is nominally set to 60 K, the stage can accommodate a temperature between 60 K and 80 K. This insight highlights the importance of considering the entire range of available temperatures to optimize the placement and operation of different stages within a cryogenic system.

By utilizing the available temperature range in each cryocooler stage, the overall performance of the computing system can be significantly enhanced. The methodology proposed in Chapter 5 can be used to determine the optimal operating temperature for each component within a cryogenic system, minimizing total power consumption while ensuring that the performance satisfies a specified constraint. This optimization process considers the individual thermal characteristics of each cryocooler stage, resulting in a more efficient distribution of components based on local thermal and power requirements. The methodology described in Chapter 5, however, does not consider nested refrigerator stages, relies on a fixed number of individual refrigerators, and individually optimizes the operation within each temperature zone rather than optimizing the overall system. Furthermore, only a fixed number of units within a system is considered.

In Chapter 6, exploration of the optimization methodology is further extended to incorporate the mutual dependence of the different temperature zones to enhance the energy efficiency of the overall system. This approach represents a transition from individually optimizing each temperature zone to a more comprehensive strategy aimed at lowering the overall system-wide energy. While the methodology described in Chapter 6 supports several refrigerator stages, the optimal temperature of each stage is determined for a pre-set number of refrigerator stages. Furthermore, the unit assigned to each refrigerator chamber is fixed, and groups of these units are not allowed. The methodology proposed in this work corrects this significant limitation by determining the optimal number of refrigerator stages. Furthermore, the algorithm groups the units according to the proposed number of refrigerator stages, allowing for more flexible and efficient temperature management across multiple units and temperature zones.

7.1.2 Applications

Superconductive computing systems are being developed as a promising alternative to traditional CMOS technology for high performance computing (HPC) [213, 215]. These systems utilize superconductive single flux quantum logic families based on Josephson junctions (JJ) [194, 213] as the active component, as opposed to MOSFETs in traditional CMOS circuits. Superconductive computing systems operate at cryogenic temperatures [191], exploiting the zero resistance of the interconnect to achieve extremely fast switching speeds and minimal power consumption [34, 215, 216]. Circuits employing passive superconductive interconnects dissipate three to four orders of magnitude less energy as opposed to CMOS interconnects [48].

Quantum computing is another technology with the potential to revolutionize a wide range of computational tasks, such as prime factorization, quantum simulation, machine learning, cryptography, and complex optimization [176, 177]. To maintain quantum coherence and minimize thermal noise, operation of quantum processors requires temperatures from a few tens to a few hundreds of millikelvins [201]. Specialized circuitry with low heat load for the quantum processors, such as readout and control of the qubits, is necessary [65, 203, 204]. These control and readout circuits typically operate at different temperature zones, from tens to hundreds of millikelvins to room temperature. This multi-temperature configuration suggests the need to optimize the temperatures across a variety of functions and refrigeration chambers.

An alternative for next generation computing is cloud computing. Given the stationary nature of cloud computing data centers, incorporating cryogenic cooling can significantly enhance performance and energy efficiency. The power consumption of individual units is significantly lower at cryogenic temperatures. Certain workloads exhibit a hundredfold reduction in power consumption at 77 K as compared to room temperature operation [46]. For instance, a CMOS-based cache operating at 77 K has been reported to consume up to ten times less energy than operating at room temperature on the same workload [47]. Similarly, DRAM at 77 K operates 3.8 times faster while consuming only 9.2% of the power required at room temperature [46]. SFQ circuits consume three to five orders of magnitude less power than CMOS operating at low temperatures [48].

7.2 Methodology

Optimization of multi-temperature zone systems is achieved here through a graph theoretic approach [178]. The proposed methodology is divided into two primary phases: the construction of the graph of the system and determining the optimal path within the graph. In the first phase, the system, composed of multi-temperature zones, is divided into multiple units or groups of units, and the corresponding graph is constructed. A unit here represents an independent functional component of the system, characterized by a power and delay which varies according to the operating temperature. As depicted in Fig. 7.2, each node (A, B, and C) in the graph represents a group of units, while the edge (T_{ij}, p_{ij}, d_{ij}) represents an operating temperature T_{ij} for a specific group. Each edge is assigned two weights: power consumption p_{ij} and delay d_{ij} at a given temperature. The optimal path through the graph corresponds to the path that minimizes the total power consumption, defined as the sum of the power weights along a path, while satisfying a target delay constraint, which is the sum of the delay weights along a path. The power weight is a cost function which depends upon the temperature of the current and previous chambers and the power consumption of the unit, as described in the next subsection. In the second phase, the optimal path is determined using several optimization techniques, including graph pruning and dynamic programming (ϵ-dominance based partitioning [217]), to accelerate the search process. After determining the optimal path for all of the constructed graphs, the results are compared to determine the most optimal number of temperature zones.

To solve the problem of temperature optimization, however, a precise mathematical formulation is first established. A formulation of the temperature optimization problem is provided in Section 7.2.1. An algorithm is proposed to solve this problem in Section 7.2.2. An assessment of the complexity of the algorithm is described in Section 7.2.3, and a method to accelerate the algorithm is described in Section 7.2.4.

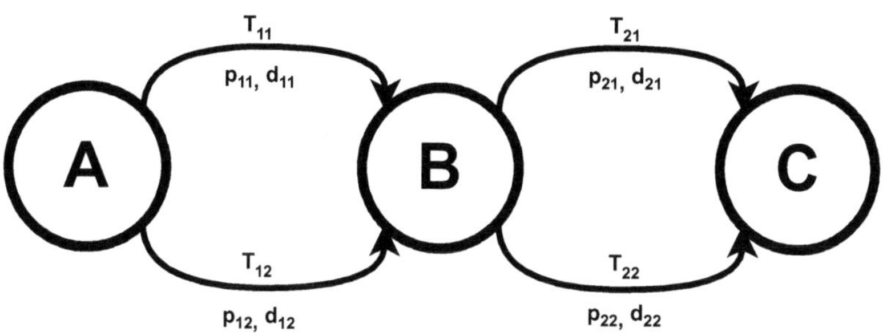

Fig. 7.2 An example of a constructed graph of a partitioned multi-temperature computing system. The nodes represent the partitioned groups of units, and the edges represent the temperature of operation and corresponding power consumption and delay of the unit operating at that temperature

7.2.1 Problem Formulation

A problem formulation is required to describe the problem and establish the necessary variables, assumptions, and objective functions. The formulation is intended to establish a foundation for developing solutions to this problem and to facilitate future improvements. A set of units, each capable of operating at different temperatures, is considered. Each unit has an associated power consumption and delay at each temperature. The objective is to determine the optimal sequence (or order) of groups, where each group consists of one or more units operating at the same temperature. Additionally, the corresponding temperature of each chamber must be identified to minimize the total power dissipation, including the power to cool the refrigeration chambers. The optimization must also ensure that the delay satisfies a target constraint.

Assumptions:

- Each group operates at a single temperature and is treated as a single unit in the optimization process. The power and delay of the group are the sum of, respectively, the power and delay of all of the individual units within that group.
- Each refrigerator chamber is located inside another refrigerator chamber to minimize power consumption and reduce wiring complexity. This nesting arrangement means that the order of the chambers (or groups) determines the temperature hierarchy, with the inner chamber operating at a lower temperature than the outer chamber.
- The units within the groups and order of the groups are not fixed. Any combination of units can form a group. For a given partition of units into k groups, there are $k!$ possible combinations of the group order.

Given:

- N units: $U_1, U_2, \ldots, U_N$
- Groups $G_1, G_2, \ldots, G_k$, where each group consists of one or more units
- Each group G_i can operate at a set of temperatures $T_i^1, T_i^2, \ldots, T_i^{n_i}$, where n_i is the number of available temperatures for group G_i.
- $P_i(T_i^m)$: The power consumption of group G_i at temperature T_i^m. If a group consists of multiple units, the power consumption of the group is the sum of the power consumption of each unit within the group.
- $D_i(T_i^m)$: The delay associated with group G_i at temperature T_i^m. If a group consists of multiple units, the delay of the group is the sum of the delay of each unit within the group.
- $C\left(P_i\left(T_i^m\right), T_i^m, T_j^n\right)$: A cost function for refrigerating a group at temperature T_i^m, while the temperature of the previous refrigerator chamber is T_j^n.

Decision variables:

- To select the temperature of each group G_i, a binary decision variable $x_{i,m} \in \{0, 1\}$ is defined, where

$$x_{i,m} = 1, \text{ if } G_i \text{ operates at temperature } T_i^m, \tag{7.1}$$

$$x_{i,m} = 0, \text{ otherwise.} \tag{7.2}$$

- To identify the sequence of groups, decision variable $y_{i,j} \in \{0, 1\}$ is defined, where

$$y_{i,j} = 1, \text{ if } G_i \text{ followed by } G_j, \tag{7.3}$$

$$y_{i,j} = 0, \text{ otherwise.} \tag{7.4}$$

The objective of the temperature optimization problem is to minimize the total power weight, which is the sum of the power weights along the path. The power weight of an edge is the cost of refrigerating a group. The objective function is therefore

$$\text{Minimize} \sum_{i=1}^{k} \sum_{j=i+1}^{k} \sum_{m=1}^{n_i} \sum_{n=1}^{n_j} C\left(P_i(T_i^m), T_i^m, T_j^n\right) \cdot x_{i,m} \cdot x_{j,n} \cdot y_{i,j}. \tag{7.5}$$

Multiple constraints exist while minimizing the objective function:

1. Each group must operate at only one temperature:

$$\sum_{m=1}^{n_i} x_{i,m} = 1, \forall i \in \{0, 1, \ldots, k\}. \tag{7.6}$$

2. To place one refrigerator chamber inside another chamber, the temperature of the inner chamber must be strictly less than the temperature of the outer chamber:

$$\text{if } \quad T_i^m < T_j^n, \quad \text{then}$$

$$x_{i,m} \cdot x_{j,n} \cdot y_{i,j} \leq 0, \forall i, j \in \{0, 1, \ldots, k\}, \tag{7.7}$$

which ensures that $T_j^n < T_i^m$, enforcing strict inequality. (This constraint might not be necessary since, in practical applications, the optimal solution naturally satisfies this constraint.)

3. The total delay must not exceed the delay constraint D_{max}:

$$\sum_{i=1}^{k} \sum_{m=1}^{n_i} D_i \left(T_i^m\right) \cdot x_{i,m} \leq D_{max},$$

$$\forall i, j \in \{0, 1, \ldots, k\}, \tag{7.8}$$

where $D_i \left(T_i^m\right)$ is the delay of group G_i at temperature T_i^m.

7.2.2 Proposed Algorithm

The proposed algorithm systematically explores all possible groupings and sequences of units, applying edge dependent shortest path computation to determine the optimal configuration. The methodology consists of the following steps:

1. Generate all possible groupings of units.
2. For each grouping configuration, all possible combinations (permutations) of groups are generated: Since the temperature of each chamber within a refrigerator must be in descending order from room temperature to cryogenic temperatures (as per constraint 2), the order of the groups is important.
3. For each configuration of groupings and permutations, determine the optimal path using an edge dependent shortest path algorithm, as shown in Algorithm 2. The algorithm is applied to each configuration to determine the most power efficient set of temperatures while satisfying the delay constraint.
4. Select the most optimal solution: The most energy efficient paths for each configuration are compared, and the lowest power consuming configuration (and set of temperatures) are selected.

Pseudocode for steps 1, 2, and 4 of the methodology is shown in Algorithm 1. The algorithm generates all possible groupings of units and all possible permutations in which a set of groups can be ordered or arranged. The specific grouping configuration and permutation sequence are passed to Algorithm 2 to determine the shortest path.

Algorithm 2 determines the most optimal path for a specific configuration and sequence by employing dynamic programming. The algorithm stores the paths in priority queue pq and the shortest distance in the dictionary $distances$. Each new path adds and/or removes a path with a power and delay worse than the newly explored path. When a new edge is explored, a cost function for refrigerating a group, $C\left(P_G\left(T\right), T, T_{amb}\right)$, is determined. This cost function depends upon the temperature of the current chamber T and the ambient temperature T_{amb}, which is

Algorithm 1 Multi-Unit Power Optimization

Input:
$df \leftarrow$ DataFrame with units, temperatures, power, and delay values
$delay\_limit \leftarrow$ Maximum allowed total delay
$Tamb \leftarrow$ Ambient starting temperature
$temps \leftarrow$ List of available temperatures
$eff\_range \leftarrow$ Efficiency values corresponding to temperatures
Output:
$best\_power \leftarrow$ Minimum total power consumption found
$best\_delay \leftarrow$ Corresponding total delay
$best\_path \leftarrow$ Optimal sequence of temperatures and units
$best\_grouping \leftarrow$ Optimal grouping of units
Initialize $best\_power \leftarrow \infty$, $best\_delay \leftarrow \infty$, $best\_path \leftarrow$ None, $best\_grouping \leftarrow$ None
Generate all possible groupings of units: $groupings \leftarrow$ GenerateAllGroupings($df.units$)
for each grouping in $groupings$ **do**
  $df\_grouped \leftarrow$ ApplyGrouping(df, $grouping$)
  $permutations \leftarrow$ AllPermutationsOfGroups($grouping$)
  for each permutation in $permutations$ **do**
    $result \leftarrow$ EdgeDependentShortestPath($df\_grouped, permutation, delay\_limit, Tamb, temps, eff\_range$)
    if $result$ is not None **then**
      $(power, delay, path) \leftarrow result$
      if $power < best\_power$ **or** ($power == best\_power$ **and** $delay < best\_delay$) **then**
        $best\_power \leftarrow power$
        $best\_delay \leftarrow delay$
        $best\_path \leftarrow path$
        $best\_grouping \leftarrow grouping$
return $best\_power, best\_delay, best\_path, best\_grouping$

the temperature of the previous refrigerator chamber,

$$C\left(P_G\left(T\right), T, T_{amb}\right) = P_G(T)\left(\frac{T_{amb}/T - 1}{\eta_{II}/100} + 1\right), \quad (7.9)$$

where η_{II} is the second law efficiency of the refrigerator [69]. After the set of optimal temperatures is obtained for each configuration and sequence, Algorithm 1 compares the results, returning the most optimal grouping and configuration.

7.2.3 Complexity

The complexity of the proposed algorithm depends upon three factors: the number of groupings, the possible permutations of the established groups, and the efficiency

Algorithm 2 Edge Dependent Shortest Path

```
Input:
        df_grouped: Grouped DataFrame with units and power/delay at each temperature
        unit_order: Ordered list of units/groups to process
        delay_limit: Maximum allowed total delay
        T_amb: Ambient starting temperature
        temps: List of available temperatures
        eff_range: Efficiency values corresponding to temperatures
Output:
        best_power, best_delay, best_path: Optimal cumulative power, delay, and path, or
    None if no valid path exists
Initialize priority queue pq with initial states for the start unit at temperatures ≤ T_amb
Initialize distances dictionary distances
Initialize best_power ← ∞, best_delay ← ∞, best_path ← None
while pq is not empty do
    Dequeue (cum_power, cum_delay, curr_unit,
    prev_temp, path) from pq
    if cum_delay > delay_limit then
        continue
    if curr_unit is the last unit in unit_order then
        if cum_power < best_power or (cum_power = best_power and cum_delay <
    best_delay) then
            best_power ← cum_power
            best_delay ← cum_delay
            best_path ← path
        continue
    next_unit ← unit after curr_unit in unit_order
    paths ← None
    for each temp in temps where temp ≤ prev_temp do
        Compute power_weight and delay_weight for curr_unit at temp
        new_cum_power ← cum_power + power_weight
        new_cum_delay ← cum_delay + delay_weight
        if new_cum_delay > delay_limit then
            continue
        Insert new path to paths
    Filter paths using ε-dominance
    Enqueue filtered paths to pq and distances
return best_power, best_delay, best_path
```

of the path finding algorithm. The number of possible combinations of groupings is determined by the Bell number [218], which increases exponentially with the number of units. Determining the optimal path for a large number of units therefore becomes computationally expensive. Furthermore, the number of refrigerator stages is limited, since more than a certain number of chambers may reduce overall power efficiency [219]. For example, the study in [219] considers a refrigerator with seven stages. In the proposed methodology, the number of refrigerator stages is limited to ten to consider possible systems which may require additional refrigerator stages.

A second factor contributing to the algorithmic complexity is the number of permutations which is $n!$, where n is the number of groups. A third factor is the complexity of the edge dependent shortest path finding algorithm. Without graph pruning, the complexity is $O(n^k)$, where k is the number of available temperatures for each refrigerator stage. After employing graph pruning, the complexity is reduced and depends upon the power and delay of each unit and the target delay. From experiments, the complexity seems close to $O(n^{k/1.5})$.

The overall complexity of the algorithm is therefore Bell(n) $\times n! \times O(n^{k/1.5})$. For a large number of available temperatures per chamber, estimating the optimal path becomes computationally expensive due to the $n^{k/1.5}$ term. Methods for accelerating the algorithm are therefore considered.

7.2.4 ϵ-Dominance Based Partitioning

All non-dominated, Pareto optimal solutions to the problem are computationally expensive. To improve the convergence in determining the optimal path and to increase the diversity of the considered paths, the dynamic programming concept of ϵ-dominance is employed [217]. With ϵ-dominance, the objective space is partitioned into boxes (hypercubes) of size ϵ to maintain the diversity of the solutions. A single solution within each box is selected to speed up the convergence process [217, 220] and can be considered as an archiving technique.

A set of objective functions,

$$\text{f(x)} = (f_1(x), f_2(x), \ldots, f_k(x)), \tag{7.10}$$

can be partitioned into hypercubes of size ϵ_j for the jth hypercube. An identification vector $b = (b_1, b_2, \ldots, b_k)$ is a vector which identifies which hypercube the objective belongs to. For each solution x, the identification vector is defined as

$$b_j \equiv (1+\delta)^{\lfloor \log_{1+\delta} \frac{f_j(x)}{f_j^{min}} \rfloor}, \quad j = 1, 2, \ldots, k, \tag{7.11}$$

where $1 + \delta$ is the ratio of the maximum value to the minimum value within the hypercube, and f_j^{min} is the minimum value of the jth objective.

In the proposed algorithm, this ϵ-dominance technique reduces the number of delay values under consideration while maintaining the diversity of the potential solutions. Before proceeding to the next node, the existing paths are archived based on the minimum delay and ϵ. The epsilon dominance approach guarantees that the retained solutions are within ϵ of the true Pareto front in the delay dimension, where the Pareto front represents the set of non-dominated solutions which offer the best tradeoff between power and delay. Epsilon dominance reduces the number of paths that needs to be stored and evaluated. A smaller ϵ retains more paths and produces a better (closer to the optimal) solution, albeit at the cost of additional computation, while a larger ϵ improves the computational efficiency but is likely to produce a less optimal result. To produce results within 10% of the desired values, a hypercube size of 10% is used ($\delta = 0.1$). Each hypercube contains solutions with objective values differing from one another by no more than 10%.

After employing dynamic programming, the complexity is further reduced due to considering fewer paths. The complexity of the edge dependent shortest path finding algorithm becomes, $O(b^2 \times k^2 \times n)$, where b is the number of boxes,

$$b \approx \frac{1}{\delta} log \frac{f^{max}}{f^{min}}. \tag{7.12}$$

Here, f^{max} and f^{max} are, respectively, the maximum and minimum value of the objective, power consumption. The overall complexity of the algorithm is therefore $\mathrm{Bell}(n) \times n! \times O(b^2 \times k^2 \times n)$.

Alternatively, gradient-based approaches can be used to accelerate the computation of the most optimal path. To ensure that gradient-based algorithms determine the global minima, the convexity of the optimized function needs to be proven. The objective function, the power consumption of a refrigerator chamber, is

$$\begin{aligned} P_i =& (P_{i+1}(T_i, T_{i+1}) + Q_i(T_i)) \\ &\cdot \left(\frac{\frac{T_{i-1}}{T_i} - 1}{\eta_{II}(T_i, T_{i-1})/100} + 1 \right). \end{aligned} \tag{7.13}$$

Since proving (7.13) is complex, gradient-based approaches have not been used in the current methodology. The only acceleration technique utilized in the edge dependent shortest path algorithm is ϵ-dominance.

7.3 Case Study

Two abstract computing systems have been analyzed as a case study. These computing systems are support circuitry for a quantum computing system (readout and control) and a cloud computing system. Each system consists of multiple units, each unit operating at different temperatures ranging from 300 K (RT) to 3 K. The power consumption and delay of each unit are randomly assigned to resemble the range of values reported in measurements [46, 221–224]. The power and delay vary at different temperatures in a exponential manner and are modeled as

$$y = y_{min} + (y_{max} - y_{min}) \frac{1 - e^{-kT}}{1 - e^{-kT_{max}}}, \tag{7.14}$$

where y is either the delay or power, T is the temperature, and k is a coefficient modeling the rate of increase or decrease of the power or delay with temperature. The expression is a simplified model which captures the approximate exponential reduction in power and delay with temperature. The power dissipation at room temperature is randomly chosen between 5 watts and 100 watts. The delay at room temperature is randomly chosen between 50 nanoseconds and 1,000 nanoseconds. The variation in power dissipation between the minimum and maximum temperature is randomly chosen to be between five to ten times, and the delay variation between the minimum and maximum temperature is randomly assigned as five to fifteen times. The coefficient k in (7.14) is randomly assigned between 0.006 to 0.015 for each unit. These ranges are based on trends in power reductions and performance improvements observed in relevant studies of cryogenic circuits [46, 221–224].

A system in the first case study, a cloud computer, consists of six units where 50 different operating temperatures are considered for each unit. A system in the second case study, support circuitry for a quantum computer, consists of seven units and 20 different operating temperatures considered for each unit. The target delay constraint is 600 nanoseconds. Optimization results for both case studies are listed in Table 7.1 where the most optimal grouping, most optimal number and sequence of chambers, and corresponding temperature for each chamber are listed. Both six and seven unit systems are shown to operate most efficiently within a three chamber refrigerator. In non-optimized cloud computing and quantum computing systems, where each unit is placed at the lowest temperature available for that unit, the total power consumption is, respectively, 17.6 kilowatts and 20.9 kilowatts. After optimizing the temperatures using the proposed methodology, these systems consume 1.4 kilowatts and 3.4 kilowatts, achieving, respectively, almost thirteen and six times reduction in power. Two additional case studies, systems with eight and nine units, are listed in Table 7.1. The most optimal number of chambers for the eight and nine unit systems are, respectively, two and six.

7.4 Conclusions

Advancements in the computing industry are increasingly operating at cryogenic temperatures; particularly, cloud computing systems and quantum computing systems. These systems often include specialized circuitry that operates across a spectrum of temperatures–from room temperature to a few kelvin–necessitating multi-temperature configurations to minimize power consumption by managing the local operating temperature of each chamber. The proposed methodology focuses on the thermal optimization of cryogenic computing systems operating across multiple temperature zones.

Unlike previous approaches that rely on a predetermined number of refrigeration stages and fixed unit configurations, the proposed algorithm dynamically optimizes both the number of temperature zones and the grouping of functional units based on the performance and power profile of each subsystem. By employing a graph theoretic approach, each unit or group of units is represented as a node in a graph with the edges denoting different operating temperatures with assigned power and delay weights. The optimal path through this graph describes a system that

Table 7.1 Set of the most optimal grouping, most optimal number and sequence of chambers, and corresponding temperatures for six, seven, eight and nine unit systems. The considered systems are cryogenic cloud computing (CC) and quantum computing (QC) systems. Each unit has 20 to 50 different operating temperatures and the target delay constraint is 600 to 800 nanoseconds for each case study. The most optimal number of chambers for the first two systems is determined to be three. For the eight and nine unit systems the most optimal number of chambers are, respectively, two and six

Type	Delay limit D_{max}, [ns]	n units	n temp-s	n chambers	Unit sequence	Temperatures T, [K]	Delay D_i, [ns]	Power P_i, [W]
CC1	600	6	50	3	[1], [3], [2, 4, 5, 6]	[28.6, 12.3, 6.4]	599.1	1,379.0
QC1	600	7	20	3	[1], [4], [2, 3, 5, 6, 7]	[12.8, 4.9, 3.0]	595.6	3,404.4
CC2	800	8	50	2	[8], [1, 2, 3, 4, 5, 6, 7]	[3.6, 3.0]	787.7	3,895.5
QC2	750	9	30	6	[8], [1], [9], [7], [6], [2, 3, 4, 5]	[23.6, 9.1, 7.8, 6.6, 3.5, 3.0]	748.0	1,493.5

minimizes the total power consumption while satisfying performance constraints, thereby maximizing energy efficiency and reducing overall power usage and local heat loads.

Two practical case studies have been evaluated to demonstrate the methodology, involving two computing systems-one with six units (cloud computer) and another with seven units (readout and control circuit for quantum computer)–each capable of operating at temperatures ranging from 300 K to 3 K. The power consumption and delay for each unit are modeled by an exponential relationship with temperature based on empirical data. The power consumption of the six unit system is reduced from 17.6 kilowatts for a non-optimized system to 1.4 kilowatts, achieving a nearly thirteen fold reduction in power. For the seven unit system, a six fold reduction in power is demonstrated, from 20.9 kilowatts to 3.4 kilowatts. In both scenarios, the optimal number of refrigerators is three.

Chapter 8
Conclusions

The computing industry is approaching fundamental physical limits that challenge the continuation of Moore's Law, which has guided the exponential growth of computational power for decades. As transistor dimensions shrink below three nanometers, issues such as excessive heat density and quantum tunneling effects have emerged, allowing large numbers of electrons to pass through transistors when the device is off (cutoff region). These limitations necessitate the exploration of alternative technologies and methodologies to continue the advancement of computing capabilities.

Several directions exist for evolving computing technologies beyond conventional scaling, including three-dimensional integration, novel transistor structures such as FinFETs and gate-all-around (GAA) devices, specialized materials such as graphene or carbon nanotubes, and entirely different compute paradigms such as quantum computing. Among these many possibilities, the utilization of cryogenic temperatures to enhance computational performance and energy efficiency has evolved into a primary path for stationary computing systems and is the focus of this book.

Operating electronic systems at cryogenic temperatures offers several fundamental benefits. The reduction in thermal energy at low temperatures decreases phonon populations, leading to reduced phonon scattering and, consequently, increased carrier mobility in semiconductors. As a result, faster switching speeds in CMOS devices is observed, supporting higher operational frequencies. Furthermore, cryogenic temperatures significantly reduce noise in CMOS circuits, improving signal integrity. The lower noise levels are crucial in sensitive applications. The integration of cryogenic CMOS in quantum computing systems enables the scaling of quantum processors from a few hundred qubits available today to potentially tens of thousands of qubits.

In addition, metals exhibit lower resistance due to diminished electron-phonon interactions, and certain conductors enter a superconducting state, eliminating the electrical resistance. In CMOS, leakage currents become negligible at cryogenic

N. Zhuldassov, E. G. Friedman, *Cryogenic Microelectronic Systems for Ultra-Low Energy and Enhanced Performance*, https://doi.org/10.1007/978-3-031-90754-8_8

temperatures, enabling dynamic logic circuits that are infeasible at room temperature due to leakage-induced state degradation.

Multiple aspects of cryogenic computing systems are explored in this book to capitalize on these aforementioned benefits of operating at cryogenic temperatures. For instance, the potential of cryogenic CMOS circuits is discussed in this book. In particular, dynamic circuits operating at cryogenic temperatures is explored. At room temperature, dynamic logic circuits suffer from significant leakage currents that degrade the stored logic state, limiting the utility of these circuits. At cryogenic temperatures, however, leakage currents are drastically reduced, supporting reliable operation of dynamic logic circuits at close to DC to much higher frequencies. Dynamic circuits at cryogenic temperatures offer opportunities for increasing the operating speed while reducing the power consumption, as well as less area by combining logic and memory. The critical temperature and frequency beyond which dynamic circuits become more advantageous than static logic circuits are also discussed.

More general questions, such as the optimization of computing systems across multiple temperatures is a key aspect of this book. In particular, the question of which components should be located within which temperature zones using which technology to maximize computing performance and energy efficiency is described. Several graph theory-based methodologies are offered to address this issue. The components are represented as nodes in a graph, and the corresponding temperature transitions as edges. The optimization algorithms minimize the total power consumption of the system while adhering to specific performance constraints and considering interdependencies between the temperature zones and interconnections. Additionally, the proposed methodologies consider the temperature characteristics of the cooling technologies, as well as the efficiency and limitations of cryogenic coolers within the optimization process. It is demonstrated that the power consumption of the entire system can be reduced despite the cost of cooling.

Summarizing, a narrow yet significant pathway toward advancing the computing industry by leveraging cryogenic temperatures is the primary topic of this book. Cryogenic computing represents an important frontier in the quest for ever increasing computational power and energy efficiency. While only a fraction of possible scenarios for the future of computing is considered, valuable insight and tools for design and optimization of these systems are hopefully provided. The proposed methodologies offer a foundation for further exploration which could potentially shape the next generation of high performance, energy efficient computing technologies and systems. These findings underscore the potential of cryogenic computing as a viable solution to overcome many of the limitations imposed by conventional scaling of semiconductor technology. By addressing challenges related to thermal optimization and circuit/system design at low temperatures, groundwork for future development in this domain is provided. Potential directions include the integration of more advanced cooling technologies, the exploration of novel materials and device structures that exploit cryogenic operation, and the development of advanced cryogenic systems that include more efficient cryogenic cooling systems. The insights provided in this book will hopefully contribute to the broader effort to continue and even accelerate the progress of the computing industry, offering a path into a future where the existing limitations of the computing industry have been overcome by operating at cryogenic temperatures.

Bibliography

1. T. Bradshaw, B. Evans, and J. Vandore, *Cryogenics: Fundamentals, Foundations and Applications*, IOP Publishing, 2022.
2. B. T. M. Willis and C. J. Carlile, *Experimental Neutron Scattering*, Oxford University Press, 2017.
3. Y. Fan and Y. Wang, "Applications of Small-Angle X-Ray Scattering/Small-Angle Neutron Scattering and Cryogenic Transmission Electron Microscopy to Understand Self-Assembly of Surfactants," *Current Opinion in Colloid & Interface Science*, Vol. 42, pp. 1–16, August 2019.
4. NASA, "Webb Innovations: Cryocooler," https://webb.nasa.gov/content/about/innovations/cryocooler.html, 2024.
5. Z. Yuan, M. Cui, Y. Xie, and C. Li, "Design and Analysis of a Small-Scale Natural Gas Liquefaction Process Adopting Single Nitrogen Expansion with Carbon Dioxide Pre-Cooling," *Applied Thermal Engineering*, Vol. 64, No. 1-2, pp. 139–146, March 2014.
6. A. Dzido, P. Krawczyk, M. Wołowicz, and K. Badyda, "Comparison of Advanced Air Liquefaction Systems in Liquid Air Energy Storage Applications," *Renewable Energy*, Vol. 184, pp. 727–739, January 2022.
7. D. Delikaris and L. Tavian, "The LHC Cryogenic System and Operational Experience from the First Three Years Run," *Journal of Cryogenics and Superconductivity Society of Japan*, Vol. 49, pp. 590–600, December 2014.
8. K. Okuno, A. Shikov, and N. Koizumi, "Superconducting Magnet System in a Fusion Reactor," *Journal of Nuclear Materials*, Vol. 329, pp. 141–147, August 2004.
9. A. R. Jha, *Superconductor Technology: Applications to Microwave, Electro-Optics, Electrical Machines, and Propulsion Sytems*, John Wiley & Sons, 1998.
10. A. R. Jha, *Infrared Technology: Applications to Electro-Optics, Photonic Devices and Sensors*, John Wiley & Sons, 2000.
11. W. F. Clay, M. Faraday, and T. Northmore, *The Liquefaction of Gases*, Edinburgh: William F. Clay, 18 Teviot Place, 1896.
12. T. O. Sloane, *Liquid Air, and the Liquefaction of Gases: Theory, History, Biography, Practical Applications, Manufacture*, Henley, 1900.
13. J. P. Joule and W. Thomson, "XV. On the Thermal Effects of Fluids in Motion-Part II," *Philosophical Transactions of the Royal Society of London*, No. 144, pp. 321–364, December 1854.
14. R. G. Scurlock, "Taming the Gases," *Scientific American*, Vol. 270, No. 3, pp. 116–118, March 1992.

N. Zhuldassov, E. G. Friedman, *Cryogenic Microelectronic Systems for Ultra-Low Energy and Enhanced Performance*, https://doi.org/10.1007/978-3-031-90754-8

15. Z. Zhao and C. Wang, *Cryogenic Engineering and Technologies: Principles and Applications of Cryogen-Free Systems*, CRC Press, 2019.
16. T. Andrews, "The Bakerian Lecture: On the Continuity of the Gaseous and Liquid States of Matter," *Philosophical Transactions of the Royal Society of London*, Vol. 159, pp. 575–590, January 1869.
17. R. Radebaugh, "Advances in Cryocoolers," *Proceedings of the International Cryogenic Engineering Conference/International Cryogenic Materials Conference*, Vol. 16, pp. 33–44, Elsevier Science, December 1997.
18. F. Papanelopoulou, "Louis Paul Cailletet: The Liquefaction of Oxygen and the Emergence of Low-Temperature Research," *Notes and Records of the Royal Society*, Vol. 67, No. 4, pp. 355–373, December 2013.
19. R. Pictet, "The Liquefaction of Oxygen and Hydrogen, and the Solidifying of Hydrogen," *The Nineteenth Century and After: a Monthly Review*, Vol. 3, No. 15, pp. 969–977, May 1878.
20. Z. Wróblewski and K. S. Olszewski, "Sur la Liquefaction de L'Azote," *Notes and Records of the Royal Society*, Vol. 96, pp. 1225–1226, April 1883.
21. R. Barron, "Liquefaction Cycles for Cryogens," *Proceedings of the Advances in Cryogenic Engineering: A Collection of Invited Papers and Contributed Papers Presented at National Technical Meetings During 1970 and 1971*, pp. 20–36, Springer, September 1972.
22. G. Claude, *Liquid Air, Oxygen, Nitrogen*, P. Blakiston's Son & Company, 1913.
23. B. Seeber, *Handbook of Applied Superconductivity, Volume 2*, Vol. 2, CRC press, 1998.
24. J. Dewar, "Liquid Hydrogen," *Science*, Vol. 8, No. 183, pp. 3–6, July 1898.
25. H. K. Onnes, "Further Experiments with Liquid Helium," *Proceedings of the Royal Netherlands Academy of Arts and Sciences (Koninklijke Nederlandse Akademie van Wetenschappen)*, Vol. 13, pp. 1093–1113, February 1911.
26. A. R. Jha, *Cryogenic Technology and Applications*, Elsevier, 2011.
27. W. Gifford and H. McMahon, "A New Refrigeration Process," *Proceedings of the International Congress of Refrigeration*, Vol. 1, pp. 354–372, August 1959.
28. H. McMahon and W. Gifford, "A New Low-Temperature Gas Expansion Cycle: Part I," *Proceedings of the Cryogenic Engineering Conference*, pp. 354–367, Springer, September 1960.
29. W. Gifford and H. McMahon, "A New Low-Temperature Gas Expansion Cycle: Part II," *Proceedings of the Cryogenic Engineering Conference*, pp. 368–372, Springer, September 1960.
30. M. Z. Getie *et al.*, "Reversed Regenerative Stirling Cycle Machine for Refrigeration Application: A Review," *International Journal of Refrigeration*, Vol. 118, pp. 173–187, October 2020.
31. J. Kohler and C. Jonkers, "Construction of a Gas Refrigerating Machine," *Philips Technical Review*, Vol. 16, No. 5, pp. 105–115, October 1954.
32. R. Radebaugh, "Historical Summary of Cryogenic Activity Prior to 1950," *Proceedings of the Cryogenic Engineering*, pp. 3–27, Springer, April 2007.
33. G. Krylov and E. G. Friedman, *Single Flux Quantum Integrated Circuit Design*, Springer, 2022.
34. N. Zhuldassov and E. G. Friedman, "Cryogenic Dynamic Logic," *Proceedings of the IEEE International Symposium on Circuits and Systems*, pp. 1–5, October 2020.
35. N. Zhuldassov, R. Bairamkulov, and E. G. Friedman, "Thermal Optimization of Hybrid Cryogenic Computing Systems," *IEEE Transactions on Very Large Scale Integration (VLSI) Systems*, Vol. 31, No. 9, pp. 1339–1346, September 2023.
36. N. Zhuldassov and E. G. Friedman, "Temperature–Frequency Boundary of Cryogenic Dynamic Logic," *Microelectronics Journal*, Vol. 135, Article 105763, March 2023.
37. N. Zhuldassov, R. Bairamkulov, and E. G. Friedman, "Heat Load Efficiency in Multi-Temperature Cryogenic Computing Systems," *Cryogenics*, Vol. 146, Article 104013, March 2025.
38. T. Jabbari and E. G. Friedman, "SFQ/DQFP Interface Circuits," *IEEE Transactions on Applied Superconductivity*, Vol. 33, No. 5, Article 1303705, August 2023.

39. W. E. Gifford and R. Longsworth, "Pulse-Tube Refrigeration," *Journal of Engineering for Industry*, Vol. 86, No. 3, pp. 264–268, August 1964.
40. R. Radebaugh, "A Review of Pulse Tube Refrigeration," *Advances in Cryogenic Engineering*, Vol. 5, Springer, pp. 1191–1205, 1990.
41. R. Buyya, J. Broberg, and A. M. Goscinski, *Cloud Computing: Principles and Paradigms*, John Wiley & Sons, 2010.
42. R. Uhlig *et al.*, "Intel Virtualization Technology," *Computer*, Vol. 38, No. 5, pp. 48–56, May 2005.
43. S. Carnot, "Reflections on the Motive Power of Fire, and on Machines Fitted to Develop That Power," *Paris: Bachelier*, Vol. 108, 1824.
44. D. A. Cardwell, D. C. Larbalestier, and A. Braginski, *Handbook of Superconductivity: Processing and Cryogenics, Volume Two*, CRC Press, 2022.
45. D. S. Holmes, A. L. Ripple, and M. A. Manheimer, "Energy-Efficient Superconducting Computing–Power Budgets and Requirements," *IEEE Transactions on Applied Superconductivity*, Vol. 23, No. 3, Article 1701610, June 2013.
46. G.-H. Lee, D. Min, I. Byun, and J. Kim, "Cryogenic Computer Architecture Modeling With Memory-Side Case Studies," *Proceedings of the ACM/IEEE Annual International Symposium on Computer Architecture*, pp. 774–787, June 2019.
47. D. Min *et al.*, "CryoCache: A Fast, Large, and Cost-Effective Cache Architecture for Cryogenic Computing," *Proceedings of the International Conference on Architectural Support for Programming Languages and Operating Systems*, pp. 449–464, March 2020.
48. T. Jabbari and E. G. Friedman, "Global Interconnects in VLSI Complexity Single Flux Quantum Systems," *Proceedings of the ACM/IEEE International Workshop on System-Level Interconnect: Problems and Pathfinding Workshop*, pp. 1–7, November 2020.
49. M. Koot and F. Wijnhoven, "Usage Impact on Data Center Electricity Needs: a System Dynamic Forecasting Model," *Applied Energy*, Vol. 291, Article 116798, June 2021.
50. A. Nordlund, M. Harrison, and J. Gess, "Viability of Cryogenic Cooling to Reduce Processor Power Consumption," *Journal of Electronic Packaging*, Vol. 143, No. 4, Article 041101, December 2021.
51. F. Ayachi, L. Yang, J. Y. Sze, and A. Romagnoli, "Cryogenic Polygeneration for Green Data Centre," *Energy Procedia*, Vol. 152, pp. 15–20, October 2018.
52. K. Ishida *et al.*, "SuperNPU: An Extremely Fast Neural Processing Unit Using Superconducting Logic Devices," *Proceedings of the Annual IEEE/ACM International Symposium on Microarchitecture*, pp. 58–72, October 2020.
53. R. P. Feynman *et al.*, "Simulating Physics with Computers," *International Journal of Theoretical Physics*, Vol. 21, No. 6/7, pp. 467–488, March 1982.
54. E. Rieffel and W. Polak, "An Introduction to Quantum Computing for Non-Physicists," *ACM Computing Surveys (CSUR)*, Vol. 32, No. 3, pp. 300–335, September 2000.
55. D. Deutsch and R. Jozsa, "Rapid Solution of Problems by Quantum Computation," *Proceedings of the Royal Society of London. Series A: Mathematical and Physical Sciences*, Vol. 439, No. 1907, pp. 553–558, December 1992.
56. D. P. DiVincenzo, "Quantum Computation," *Science*, Vol. 270, No. 5234, pp. 255–261, October 1995.
57. S. McArdle *et al.*, "Quantum Computational Chemistry," *Reviews of Modern Physics*, Vol. 92, No. 1, Article 015003, March 2020.
58. V. Lordi and J. M. Nichol, "Advances and Opportunities in Materials Science for Scalable Quantum Computing," *MRS Bulletin*, Vol. 46, pp. 589–595, July 2021.
59. B. Bauer, S. Bravyi, M. Motta, and G. K.-L. Chan, "Quantum Algorithms for Quantum Chemistry and Quantum Materials Science," *Chemical Reviews*, Vol. 120, No. 22, pp. 12 685–12 717, October 2020.
60. D. J. Egger *et al.*, "Quantum Computing for Finance: State-of-the-Art and Future Prospects," *IEEE Transactions on Quantum Engineering*, Vol. 1, No. 3101724, pp. 1–24, October 2020.

61. C. Outeiral *et al.*, "The Prospects of Quantum Computing in Computational Molecular Biology," *Wiley Interdisciplinary Reviews: Computational Molecular Science*, Vol. 11, No. 1, pp. 1–21, January 2021.
62. B. Rawat *et al.*, "Quantum Computing and AI: Impacts & Possibilities," *ADI Journal on Recent Innovation*, Vol. 3, No. 2, pp. 202–207, January 2022.
63. A. Bayerstadler *et al.*, "Industry Quantum Computing Applications," *EPJ Quantum Technology*, Vol. 8, No. 25, pp. 1–17, December 2021.
64. F. Bova, A. Goldfarb, and R. G. Melko, "Commercial Applications of Quantum Computing," *EPJ Quantum Technology*, Vol. 8, No. 2, pp. 1–13, December 2021.
65. B. Patra *et al.*, "Cryo-CMOS Circuits and Systems for Quantum Computing Applications," *IEEE Journal of Solid-State Circuits*, Vol. 53, No. 1, pp. 309–321, September 2017.
66. E. Charbon *et al.*, "Cryo-CMOS for Quantum Computing," *Proceedings of the IEEE International Electron Devices Meeting*, pp. 13–5, December 2016.
67. E. Jeffrey *et al.*, "Fast Accurate State Measurement with Superconducting Qubits," *Physical Review Letters*, Vol. 112, No. 19, Article 190504, May 2014.
68. J. Elzerman *et al.*, "Single-Shot Read-Out of an Individual Electron Spin in a Quantum Dot," *Nature*, Vol. 430, No. 6998, pp. 431–435, July 2004.
69. R. G. Ross Jr, "Refrigeration Systems for Achieving Cryogenic Temperatures," *Proceedings of the Low Temperature Materials and Mechanisms*, pp. 127–200, CRC Press, August 2016.
70. Y. A. Çengel, R. H. Turner, J. M. Cimbala, and M. Kanoglu, *Fundamentals of Thermal-Fluid Sciences*, Vol. 703, McGraw-Hill, 2001.
71. E. Brunner, "Fluid Mixtures at High Pressures VI. Phase Separation and Critical Phenomena in 18 (n-Alkane+ Ammonia) and 4 (n-Alkane+ Methanol) Mixtures," *The Journal of Chemical Thermodynamics*, Vol. 20, No. 3, pp. 273–297, March 1988.
72. V. Vesovic *et al.*, "The Transport Properties of Carbon Dioxide," *Journal of Physical and Chemical Reference Data*, Vol. 19, No. 3, pp. 763–808, May 1990.
73. P. Attard, *Entropy Beyond the Second Law: Thermodynamics and Statistical Mechanics for Equilibrium, Non-Equilibrium, Classical, and Quantum Systems*, series [IOP release 5], Bristol [England] (Temple Circus, Temple Way, Bristol BS1 6HG, UK): IOP Publishing, 2018.
74. M. Donabedian, *Spacecraft Thermal Control Handbook: Cryogenics*, Vol. 2, Article 13, American Institute of Aeronautics and Astronautics, 2002.
75. P. W. Anderson, "Considerations on the Flow of Superfluid Helium," *Reviews of Modern Physics*, Vol. 38, No. 2, Article 298, April 1966.
76. S. W. Van Sciver, K. Timmerhaus, and A. F. Clark, *Helium Cryogenics*, Springer, 2012.
77. J. S. Brooks and R. J. Donnelly, "The Calculated Thermodynamic Properties of Superfluid Helium-4," *Journal of Physical and Chemical Reference Data*, Vol. 6, No. 1, pp. 51–104, January 1977.
78. P. Selzer, W. Fairbank, and C. Everitt, "A Superfluid Plug for Space," *Proceedings of the Advances in Cryogenic Engineering*, pp. 277–281, Springer, June 1971.
79. E. W. Lemmon, I. H. Bell, M. L. Huber, and M. O. McLinden, "NIST Standard Reference Database 23: Reference Fluid Thermodynamic and Transport Properties-REFPROP, Version 10.0, National Institute of Standards and Technology," 2018.
80. E. Gerjuoy, "Helium Conservation: A Policy Matter Not to Be Taken Lightly," *Duquesne Law Review*, Vol. 17, pp. 115–132, January 1978.
81. R. Dhuley *et al.*, "First Demonstration of a Cryocooler Conduction Cooled Superconducting Radiofrequency Cavity Operating at Practical cw Accelerating Gradients," *Superconductor Science and Technology*, Vol. 33, No. 6, Article 06LT01, April 2020.
82. W. E. Gifford, "The Gifford-McMahon Cycle," *Proceedings of the Advances in Cryogenic Engineering*, pp. 152–159, Springer, August 1966.
83. R. C. Longsworth, "4 K Gifford McMahon/Joule-Thomson Cycle Refrigerators," *Proceedings of the Advances in Cryogenic Engineering*, pp. 689–698, Springer, 1988.
84. W. Little, "Recent Developments in Joule-Thomson Cooling: Gases, Coolers, and Compressors," *Proceedings of the International Cryocooler Conference*, Vol. 3, August 1988.

85. G. Brunner, "Properties of Pure Water," *Hydrothermal and Supercritical Water Processes*, Vol. 5, Elsevier, pp. 9–93, 2014.
86. D. J. Missimer, *Auto-Refrigerating Cascade (ARC) Systems: An Overview*, American Institute of Chemical Engineers, 1995.
87. R. Radebaugh, "Recent Developments in Cryocoolers," *Proceedings of the International Congress of Refrigeration*, Vol. 19, pp. 973–989, August 1995.
88. E. Marquardt, R. Radebaugh, and J. Dobak, "A Cryogenic Catheter for Treating Heart Arrhythmia," *Proceedings of the Advances in Cryogenic Engineering*, Vol. 43/A, pp. 903–910, Plenum Press, June 1998.
89. V. Kotsubo *et al.*, "Compact 2.2 K Cooling System for Superconducting Nanowire Single Photon Detectors," *IEEE Transactions on Applied Superconductivity*, Vol. 27, No. 4, pp. 1–5, January 2017.
90. J. A. McCormick, W. L. Swift, and H. Sixsmith, "Progress on the Development of Miniature Turbomachines for Low-Capacity Reverse-Brayton Cryocoolers," *Cryocoolers*, Vol. 9, Springer US, pp. 475–483, 1997.
91. W. Swift, F. Dolan, and M. Zagarola, "The NICMOS Cooling system–5 Years of Successful On-Orbit Operation," *Proceedings of the American Institute of Physics Conference*, Vol. 985, No. 1, pp. 799–806, March 2008.
92. S. C. Collins, "Helium Liquefier," *Science*, Vol. 116, No. 3012, pp. 289–294, September 1952.
93. D. Thombare and S. Verma, "Technological Development in the Stirling Cycle Engines," *Renewable and Sustainable Energy Reviews*, Vol. 12, No. 1, pp. 1–38, January 2008.
94. G. Walker and E. R. Bingham, *Low-Capacity Cryogenic Refrigeration*, Oxford University Press, 1994.
95. T. Kuriyama *et al.*, "High Efficient Two-Stage GM Refrigerator with Magnetic Material in the Liquid Helium Temperature Region," *Proceedings of the Advances in Cryogenic Engineering: Part A & B*, pp. 1261–1269, Springer, December 1990.
96. T. Ohtsuka and Y. Ishizaki, *Generation of Low Temperature and its Application*, Shonan Kamakura, 2003.
97. E. I. Mikulin, A. A. Tarasov, and M. P. Shkrebyonock, "Low-Temperature Expansion Pulse Tubes," *Proceedings of the Advances in Cryogenic Engineering*, Vol. 29, pp. 629–637, Springer, January 1984.
98. R. Radebaugh, J. Zimmerman, D. Smith, and B. Louie, "A Comparison of Three Types of Pulse Tube Refrigerators: New Methods for Reaching 60K," *Proceedings of the Advances in Cryogenic Engineering: Part A & B*, Vol. 31, pp. 779–789, Springer, August 1986.
99. R. Radebaugh, "Development of the Pulse Tube Refrigerator as an Efficient and Reliable Cryocooler," *Proceedings of the Institute of Refrigeration*, Vol. 96, pp. 11–29, January 2001.
100. E. D. Marquardt and R. Radebaugh, "A Pulse Tube Oxygen Liquefier," *Proceedings of the Advances in Cryogenic Engineering*, Vol. 45/A, pp. 457–464, Plenum Press, September 2000.
101. X. Wang *et al.*, "A High Efficiency Hybrid Stirling-Pulse Tube Cryocooler," *AIP Advances*, Vol. 5, No. 3, March 2015.
102. P. Kittel, "Ideal Orifice Pulse Tube Refrigerator Performance," *Cryogenics*, Vol. 32, No. 9, pp. 843–844, January 1992.
103. R. D. Dupuis and M. R. Krames, "History, Development, and Applications of High-Brightness Visible Light-Emitting Diodes," *Journal of Lightwave Technology*, Vol. 26, No. 9, pp. 1154–1171, March 2008.
104. E. A. Gutierrez-D, J. Deen, and C. Claeys, *Low Temperature Electronics: Physics, Devices, Circuits, and Applications*, Elsevier, 2000.
105. Y. Feng *et al.*, "Characterization and Modelling of MOSFET Operating at Cryogenic Temperature for Hybrid Superconductor-CMOS Circuits," *Semiconductor Science and Technology*, Vol. 19, No. 12, Article 1381, October 2004.
106. Y. Ukhanov and Y. Mal'tsev, "Investigation of the Temperature Dependence of the Effective Electron Mass in n-InAs in the Region 293–603°K," *Fizika Tverdogo Tela*, Vol. 5, pp. 1124–1126, December 1963.

107. K. K. Thornber, "Relation of Drift Velocity to Low-Field Mobility and High-Field Saturation Velocity," *Journal of Applied Physics*, Vol. 51, No. 4, pp. 2127–2136, April 1980.
108. K. Thornber, "Applications of Scaling to Problems in High-Field Electronic Transport," *Journal of Applied Physics*, Vol. 52, No. 1, pp. 279–290, January 1981.
109. J. Huang and J. W. Schrankler, "Switching Characteristics of Scaled CMOS Circuits at 77 K," *IEEE Transactions on Electron Devices*, Vol. 34, No. 1, pp. 101–106, January 1987.
110. S. Hanamura *et al.*, "Low-Temperature CMOS 8×8 Bit Multipliers with Sub-10-ns Speeds," *IEEE Transactions on Electron Devices*, Vol. 34, No. 1, pp. 94–100, January 1987.
111. C. C. Hu, "Hot-Electron Effects in MOSFETs," *Proceedings of the IEEE International Electron Devices Meeting*, pp. 176–181, December 1983.
112. G. Baccarani and M. Wordeman, "An Investigation of Steady-State Velocity Overshoot in Silicon," *Solid-State Electronics*, Vol. 28, No. 4, pp. 407–416, April 1985.
113. J.-H. Rhew, Z. Ren, and M. S. Lundstrom, "A Numerical Study of Ballistic Transport in a Nanoscale MOSFET," *Solid-State Electronics*, Vol. 46, No. 11, pp. 1899–1906, November 2002.
114. C. Kittel, *Introduction to Solid State Physics*, John Wiley & Sons, 2005.
115. R. P. Feynman, A. R. Hibbs, and D. F. Styer, *Quantum Mechanics and Path Integrals*, Courier Corporation, 2010.
116. S. Dhar, *Analytical Mobility Models for Strained Silicon-Based Devices*, Ph.D. Dissertation, Technische Universität Wien, 2007.
117. J. R. Chelikowsky and M. L. Cohen, "Nonlocal Pseudopotential Calculations for the Electronic Structure of Eleven Diamond and Zinc-Blende Semiconductors," *Physical Review B*, Vol. 14, No. 2, Article 556, July 1976.
118. T. B. Boykin, G. Klimeck, and F. Oyafuso, "Valence Band Effective-Mass Expressions in the $sp^3d^5s^*$ Empirical Tight-Binding Model Applied to a Si and Ge Parametrization," *Physical Review B*, Vol. 69, No. 11, Article 115201, March 2004.
119. S. Adachi, *Handbook on Physical Properties of Semiconductors*, Springer Science & Business Media, April 2004.
120. M. Asche *et al.*, *Hot-Electron Transport in Semiconductors*, Vol. 58, Springer Science & Business Media, January 2006.
121. A. F. J. Levi, *Essential Semiconductor Laser Physics*, Morgan & Claypool Publishers, 2018.
122. G. Ottaviani *et al.*, "Hole Drift Velocity in Silicon," *Physical Review B*, Vol. 12, No. 8, Article 3318, October 1975.
123. I. Balslev and P. Lawaetz, "On the Interpretation of the Observed Hole Mass Shift with Uniaxial Stress in Silicon," *Physics Letters*, Vol. 19, No. 1, pp. 6–7, January 1965.
124. B. K. Ridley, *Quantum Processes in Semiconductors*, Oxford University Press, August 2013.
125. J. A. Van Vechten, "Quantum Dielectric Theory of Electronegativity in Covalent Systems. I. Electronic Dielectric Constant," *Physical Review*, Vol. 182, No. 3, pp. 891–905, June 1969.
126. J. Van Vechten, "Quantum Dielectric Theory of Electronegativity in Covalent Systems. II. Ionization Potentials and Interband Transition Energies," *Physical Review*, Vol. 187, No. 3, pp. 1007–1020, November 1969.
127. W. Bludau, A. Onton, and W. Heinke, "Temperature Dependence of the Band Gap of Silicon," *Journal of Applied Physics*, Vol. 45, No. 4, pp. 1846–1848, April 1974.
128. R. Stradling and V. Zhukov, "Cyclotron Resonance of Electrons in Silicon at Temperatures up to 200° K," *Proceedings of the Physical Society*, Vol. 87, No. 1, pp. 263–271, January 1966.
129. M. Cardona, W. Paul, and H. Brooks, "The Temperature Dependence of the Polarizability of the Free Carriers in Germanium and Silicon," *Helvetica Physica Acta*, Vol. 33, pp. 329–346, January 1960.
130. E. Sonder and D. Stevens, "Magnetic Properties of N-Type Silicon," *Physical Review*, Vol. 110, No. 5, Article 1027, June 1958.
131. H. D. Barber, "Effective Mass and Intrinsic Concentration in Silicon," *Solid-State Electronics*, Vol. 10, No. 11, pp. 1039–1051, November 1967.

132. H. Miyazawa, K. Suzuki, and H. Maeda, "Evidence for the $< 110 >$ Swelling Constant Energy Surface for Heavy Holes in Silicon," *Physical Review*, Vol. 131, No. 6, Article 2442, September 1963.
133. W. Pauli, "Über Den Zusammenhang Des Abschlusses Der Elektronengruppen im Atom Mit Der Komplexstruktur Der Spektren," *Zeitschrift FÜR Physik*, Vol. 31, No. 1, pp. 765–783, February 1925.
134. B. I. Shklovskii and A. L. Efros, *Electronic Properties of Doped Semiconductors*, Vol. 45, Springer Science & Business Media, 2013.
135. P. A. M. Dirac, "On the Theory of Quantum Mechanics," *Proceedings of the Royal Society of London. Series A, Containing Papers of a Mathematical and Physical Character*, Vol. 112, No. 762, pp. 661–677, October 1926.
136. E. Fermi, "Sulla Quantizzazione Del Gas Perfetto Monoatomico," *Rendiconti Lincei*, Vol. 145, No. 9, 1926.
137. H. Hernandez, "Standard Maxwell-Boltzmann Distribution: Definition and Properties," *ForsChem Research Reports*, Vol. 2, Article 2017–2, February 2017.
138. C.-T. Sah *et al.*, "Effect of Zinc Impurity on Silicon Solar-Cell Efficiency," *IEEE Transactions on Electron Devices*, Vol. 28, No. 3, pp. 304–313, March 1981.
139. E. Conwell and V. Weisskopf, "Theory of Impurity Scattering in Semiconductors," *Physical Review*, Vol. 77, No. 3, Article 388, February 1950.
140. H. Brooks, "Scattering by Ionized Impurities in Semiconductors," *Proceedings of the Physical Review*, Vol. 83, pp. 879–879, January 1951.
141. D. M. Caughey and R. Thomas, "Carrier Mobilities in Silicon Empirically Related to Doping and Field," *Proceedings of the IEEE*, Vol. 55, No. 12, pp. 2192–2193, December 1967.
142. J. Dorkel and P. Leturcq, "Carrier Mobilities in Silicon Semi-Empirically Related to Temperature, Doping and Injection Level," *Solid-State Electronics*, Vol. 24, No. 9, pp. 821–825, September 1981.
143. N. Sclar, "Neutral Impurity Scattering in Semiconductors," *Physical Review*, Vol. 104, No. 6, Article 1559, December 1956.
144. K. C. Kwong, *Neutral Impurity Scattering in Semiconductors*, Ph.D. Dissertation, Louisiana State University, 1990.
145. C. Erginsoy, "Neutral Impurity Scattering in Semiconductors," *Physical Review*, Vol. 79, No. 6, pp. 1013–1014, September 1950.
146. S. S. Li and W. R. Thurber, "The Dopant Density and Temperature Dependence of Electron Mobility and Resistivity in n-Type Silicon," *Solid-State Electronics*, Vol. 20, No. 7, pp. 609–616, July 1977.
147. J. Bass, "Deviations from Matthiessen's Rule," *Advances in Physics*, Vol. 21, No. 91, pp. 431–604, May 1972.
148. J. Singh, *Electronic and Optoelectronic Properties of Semiconductor Structures*, Cambridge University Press, 2007.
149. S. Selberherr, W. Hänsch, M. Seavey, and J. Slotboom, "The Evolution of the MINIMOS Mobility Model," *Solid-State Electronics*, Vol. 33, No. 11, pp. 1425–1436, November 1990.
150. A. P. French, *An Introduction to Quantum Physics*, Routledge, 2018.
151. C. Jacoboni, C. Canali, G. Ottaviani, and A. A. Quaranta, "A Review of Some Charge Transport Properties of Silicon," *Solid-State Electronics*, Vol. 20, No. 2, pp. 77–89, February 1977.
152. J. P. Uyemura, *CMOS Logic Circuit Design*, Springer, 2002.
153. A. Akturk *et al.*, "Device Modeling at Cryogenic Temperatures: Effects of Incomplete Ionization," *IEEE Transactions on Electron Devices*, Vol. 54, No. 11, pp. 2984–2990, October 2007.
154. A. Van Der Ziel, "Thermal Noise in Field-Effect Transistors," *Proceedings of the IRE*, Vol. 50, No. 8, pp. 1808–1812, August 1962.
155. Y. Feng *et al.*, "Josephson-CMOS Hybrid Memory With Ultra-High-Speed Interface Circuit," *IEEE Transactions on Applied Superconductivity*, Vol. 13, No. 2, pp. 467–470, July 2003.
156. L. Wanhammar, *DSP Integrated Circuits*, Elsevier, 1999.

157. H. Homulle, L. Song, E. Charbon, and F. Sebastiano, "The Cryogenic Temperature Behavior of Bipolar, MOS, and DTMOS Transistors in Standard CMOS," *IEEE Journal of the Electron Devices Society*, Vol. 6, pp. 263–270, January 2018.
158. A. Jonscher, "Semiconductors at Cryogenic Temperatures," *Proceedings of the IEEE*, Vol. 52, No. 10, pp. 1092–1104, October 1964.
159. R. W. Keyes, E. P. Harris, and K. Konnerth, "The Role of Low Temperatures in the Operation of Logic Circuitry," *Proceedings of the IEEE*, Vol. 58, No. 12, pp. 1914–1932, December 1970.
160. G. B. Craig and S. S. Sesnic, "Investigations of Field Effect Transistors at Cryogenic Temperatures," Texas University Austin Electronics Research Center, Technical Report, January 1970.
161. S. Sesnic and G. Craig, "Thermal Effects in JFET and MOSFET Devices at Cryogenic Temperatures," *IEEE Transactions on Electron Devices*, Vol. 19, No. 8, pp. 933–942, August 1972.
162. A. R. Tokuda and P. Lauritzen, "MOSFET Thresholds at 4.2 K Induced by Cooling Bias," *IEEE Transactions on Electron Devices*, Vol. 21, No. 9, pp. 606–607, September 1974.
163. B. Lengeler, "Semiconductor Devices Suitable for Use in Cryogenic Environments," *Cryogenics*, Vol. 14, No. 8, pp. 439–447, August 1974.
164. R. L. Maddox, "P-MOSFET Parameters at Cryogenic Temperatures," *IEEE Transactions on Electron Devices*, Vol. 23, No. 1, pp. 16–21, January 1976.
165. J. C. Whitaker, *The Electronics Handbook*, CRC Press, 2005.
166. R. M. Incandela *et al.*, "Characterization and Compact Modeling of Nanometer CMOS Transistors at Deep-Cryogenic Temperatures," *IEEE Journal of the Electron Devices Society*, Vol. 6, pp. 996–1006, April 2018.
167. Z. Chen *et al.*, "Temperature Dependences of Threshold Voltage and Drain-Induced Barrier Lowering in 60 nm Gate Length MOS Transistors," *Microelectronics Reliability*, Vol. 54, No. 6-7, pp. 1109–1114, June 2014.
168. S. Garg and T. K. Gupta, "Low Power Domino Logic Circuits in Deep-Submicron Technology Using CMOS," *Engineering Science and Technology, an International Journal*, Vol. 21, No. 4, pp. 625–638, August 2018.
169. P. Sharma *et al.*, "Design and Operational Analysis of a Green Data Center," *IEEE Internet Computing*, Vol. 21, No. 4, pp. 16–24, August 2017.
170. B. Oy, (2023) Technical specifications of BlueFors Oy cryocoolers, [Online]. Available: https://bluefors.com/products
171. L. Sumitomo Heavy Industries, (2023) Technical specifications of Sumitomo cryocoolers, [Online]. Available: https://www.shi.co.jp/english/products/machinery/cold/index.html
172. S. Inc, (2023) Technical specifications of Sunpower cryocoolers, [Online]. Available: https://www.sunpowerinc.com/products/stirling-cryocoolers/cryocoolers-overview
173. N. Grumman, (2023) Technical specifications of Northrop Grumman cryocoolers, [Online]. Available: https://www.northropgrumman.com/space/cryocoolers/
174. O. A. Mukhanov *et al.*, "Superconductor Digital-RF Receiver Systems," *IEICE Transactions on Electronics*, Vol. 91, No. 3, pp. 306–317, March 2008.
175. D. Gupta *et al.*, "Modular, Multi-Function Digital-RF Receiver Systems," *IEEE Transactions on Applied Superconductivity*, Vol. 21, No. 3, pp. 883–890, December 2010.
176. L. Vandersypen and A. van Leeuwenhoek, "Quantum Computing–the Next Challenge in Circuit and System Design," *Proceedings of the IEEE International Solid-State Circuits Conference*, pp. 24–29, February 2017.
177. A. Montanaro, "Quantum Algorithms: an Overview," *npj Quantum Information*, Vol. 2, No. 1, pp. 1–8, January 2016.
178. R. Bairamkulov and E. G. Friedman, *Graphs in VLSI*, Springer, 2023.
179. G. B. Mathews, "On the Partition of Numbers," *Proceedings of the London Mathematical Society*, Vol. 1, No. 1, pp. 486–490, 1896.
180. S. Martello and P. Toth, "Solution of the Zero-One Multiple Knapsack Problem," *European Journal of Operational Research*, Vol. 4, No. 4, pp. 276–283, April 1980.

181. K. Chowdhury and S. Sarangi, "Performance of Cryogenic Heat Exchangers with Heat Leak from the Surroundings," *Advances in Cryogenic Engineering*, Vol. 29, Springer, pp. 273–280, 1984.
182. D. W. Hahn and M. N. Özisik, *Heat Conduction*, John Wiley & Sons, 2012.
183. H.-K. Lyeo and D. G. Cahill, "Thermal Conductance of Interfaces Between Highly Dissimilar Materials," *Physical Review B*, Vol. 73, No. 14, p. 144301, April 2006.
184. D. Thornburg, E. Thall, and J. Brous, "A Manual of Materials for Microwave Tubes," Radio Corporation of America, Technical Report, January 1961.
185. C. Schmidt, "Simple Method to Measure the Thermal Conductivity of Technical Superconductors, e.g., NbTi," *Review of Scientific Instruments*, Vol. 50, No. 4, pp. 454–457, April 1979.
186. P. E. Bradley and R. Radebaugh, "Properties of Selected Materials at Cryogenic Temperatures," *National Institute of Standards and Technology*, Vol. 680, pp. 1–14, June 2013.
187. N. Simon, E. Drexler, and R. Reed, "Properties of Copper and Copper Alloys at Cryogenic Temperatures. Final Report," National Institute of Standards and Technology (MSEL), Boulder, CO (United States), Technical Report, February 1992.
188. A. G. Fowler, M. Mariantoni, J. M. Martinis, and A. N. Cleland, "Surface Codes: Towards Practical Large-Scale Quantum Computation," *Physical Review A*, Vol. 86, No. 3, pp. 032 324.1–032 324.48, September 2012.
189. L. Song, H. Homulle, E. Charbon, and F. Sebastiano, "Characterization of Bipolar Transistors for Cryogenic Temperature Sensors in Standard CMOS," *Proceedings of the IEEE SENSORS Conference*, pp. 1–3, October 2016.
190. I. Conway Lamb *et al.*, "An FPGA-Based Instrumentation Platform for Use at Deep Cryogenic Temperatures," *Review of Scientific Instruments*, Vol. 87, No. 1, p. 014701, January 2016.
191. A. Mitrovic and E. G. Friedman, "Thermal Modeling of Rapid Single Flux Quantum Circuit Structures," *IEEE Transactions on Electron Devices*, Vol. 69, No. 5, pp. 2718–2724, May 2022.
192. A. Mitrovic and E. G. Friedman, "Thermal Exploration of RSFQ Integrated Circuits," *IEEE Transactions on Very Large Scale Integration (VLSI) Systems*, Vol. 32, No. 4, pp. 728–738, April 2024.
193. V. Lacquaniti *et al.*, "Engineering Overdamped Niobium-Based Josephson Junctions for Operation Above 4.2 K," *IEEE Transactions on Applied Superconductivity*, Vol. 19, No. 3, pp. 234–237, June 2009.
194. T. Jabbari, M. Bocko, and E. G. Friedman, "All-JJ Logic Based on Bistable JJs," *IEEE Transactions on Applied Superconductivity*, Vol. 33, No. 5, Article 1303807, August 2023.
195. N. Takeuchi, D. Ozawa, Y. Yamanashi, and N. Yoshikawa, "An Adiabatic Quantum Flux Parametron as an Ultra-Low-Power Logic Device," *Superconductor Science and Technology*, Vol. 26, No. 3, p. 035010, January 2013.
196. V. K. Semenov, Y. A. Polyakov, and S. K. Tolpygo, "New AC-Powered SFQ Digital Circuits," *IEEE Transactions on Applied Superconductivity*, Vol. 25, No. 3, pp. 1–7, June 2014.
197. "IARPA SuperTools Program," 2018, https://www.iarpa.gov/index.php/research-programs/supertools, Last accessed on May 2022.
198. G. Krylov, J. Kawa, and E. G. Friedman, "Design Automation of Superconductive Digital Circuits: A Review." *IEEE Nanotechnology Magazine*, Vol. 15, No. 6, pp. 54–67, December 2021.
199. Y. Ando *et al.*, "Design and Demonstration of an 8-bit Bit-Serial RSFQ Microprocessor: CORE e4," *IEEE Transactions on Applied Superconductivity*, Vol. 26, No. 5, pp. 1–5, August 2016.
200. A. G. Qoutb and E. G. Friedman, "MTJ Magnetization Switching Mechanisms for IoT Applications," *Proceedings of the Great Lakes Symposium on VLSI*, pp. 347–352, May 2018.
201. M. H. Devoret, A. Wallraff, and J. M. Martinis, "Superconducting Qubits: A Short Review," *arXiv preprint cond-mat/0411174*, November 2004.

202. M. Reiher *et al.*, "Elucidating Reaction Mechanisms on Quantum Computers," *Proceedings of the National Academy of Sciences*, Vol. 114, No. 29, pp. 7555–7560, July 2017.
203. O. Mukhanov *et al.*, "Scalable Quantum Computing Infrastructure Based on Superconducting Electronics," *Proceedings of the IEEE International Electron Devices Meeting*, pp. 31.2.1–31.2.4, December 2019.
204. N. K. Katam, O. A. Mukhanov, and M. Pedram, "Superconducting Magnetic Field Programmable Gate Array," *IEEE Transactions on Applied Superconductivity*, Vol. 28, No. 2, pp. 1–12, January 2018.
205. C. Y. Ho, R. W. Powell, and P. E. Liley, "Thermal Conductivity of the Elements," *Journal of Physical and Chemical Reference Data*, Vol. 1, No. 2, pp. 279–421, April 1972.
206. S. A. Dyer and J. S. Dyer, "Cubic-Spline Interpolation: Part 1," *IEEE Instrumentation & Measurement Magazine*, Vol. 4, No. 1, pp. 44–46, March 2001.
207. Cryomech, (2023) Technical specifications of Cryomech's Gifford-McMahon cryocoolers, [Online]. Available: https://www.cryomech.com/cryocoolers/gifford-mcmahon-cryocoolers/
208. I. Lake Shore Cryotronics, (2023) Technical specifications of ST-500 cryostat, [Online]. Available: https://www.lakeshore.com/products/product-detail/janis/st-500-microscopy-cryostat
209. S. C. Group, (2023) Cryocooler product catalogue of SHI Cryogenics Group, [Online]. Available: https://www.shicryogenics.com/wp-content/uploads/2020/10/Cryocooler-Product-Catalogue-Letter-06.20-Web.pdf
210. J. Paasschens, S. Harmsma, and R. Van der Toorn, "Dependence of Thermal Resistance on Ambient and Actual Temperature," *Proceedings of the Bipolar/BiCMOS Circuits and Technology Meeting*, pp. 96–99, September 2004.
211. CryoCoax, (2023) Technical Specifications of CryoCoax's Cryogenic Cables, [Online]. Available: https://cryocoax.com/cryogenic-cable-and-cable-assemblies/
212. M. Arvindhan, "Effective Motivational Factors and Comprehensive Study of Information Security and Policy Challenges," *System Assurances: Modeling and Management*, Chapter 29, Elsevier, pp. 531–545, 2022.
213. G. Krylov, T. Jabbari, and E. G. Friedman, *Single Flux Quantum Integrated Circuit Design, Second Edition*, Springer, 2024.
214. T. Jabbari, G. Krylov, S. Whiteley, J. Kawa, and E. G. Friedman, "Repeater Insertion in SFQ Interconnect," *IEEE Transactions on Applied Superconductivity*, Vol. 30, No. 8, Article 5400508, December 2020.
215. J. Huang, R. Fu, X. Ye, and D. Fan, "A Survey on Superconducting Computing Technology: Circuits, Architectures and Design Tools," *CCF Transactions on High Performance Computing*, Vol. 4, No. 1, pp. 1–22, March 2022.
216. T. Jabbari, G. Krylov, S. Whiteley, E. Mlinar, J. Kawa, and E. G. Friedman, "Interconnect Routing for Large Scale RSFQ Circuits," *IEEE Transactions on Applied Superconductivity*, Vol. 29, No. 5, Article 1112805, August 2019.
217. M. Laumanns, L. Thiele, K. Deb, and E. Zitzler, "Combining Convergence and Diversity in Evolutionary Multiobjective Optimization," *Evolutionary Computation*, Vol. 10, No. 3, pp. 263–282, September 2002.
218. M. Aigner, "A Characterization of the Bell Numbers," *Discrete Mathematics*, Vol. 205, No. 1-3, pp. 207–210, July 1999.
219. R. J. Thomas, P. Ghosh, and K. Chowdhury, "Optimum Number of Stages and Intermediate Pressure Level for Highest Exergy Efficiency in Large Helium Liquefiers," *International Journal of Refrigeration*, Vol. 36, No. 8, pp. 2438–2457, June 2013.
220. A. Menchaca-Méndez *et al.*, "A Co-Evolutionary Scheme for Multi-Objective Evolutionary Algorithms Based on ϵ-Dominance," *IEEE Access*, Vol. 7, pp. 18 267–18 283, February 2019.
221. I. Byun *et al.*, "A Next-Generation Cryogenic Processor Architecture," *IEEE Micro*, Vol. 41, No. 3, pp. 80–86, March 2021.

222. I. Byun *et al.*, "CryoCore: A Fast and Dense Processor Architecture for Cryogenic Computing," *Proceedings of the IEEE Annual International Symposium on Computer Architecture*, pp. 335–348, May 2020.
223. E. Garzón, A. Teman, and M. Lanuzza, "Embedded Memories for Cryogenic Applications," *Electronics*, Vol. 11, No. 1, p. 61, December 2021.
224. S. Resch, H. Cilasun, and U. R. Karpuzcu, "Cryogenic PIM: Challenges & Opportunities," *IEEE Computer Architecture Letters*, Vol. 20, No. 1, pp. 74–77, May 2021.

Index

N. Zhuldassov, E. G. Friedman, *Cryogenic Microelectronic Systems for Ultra-Low Energy and Enhanced Performance*, https://doi.org/10.1007/978-3-031-90754-8

MIX
Papier aus verantwortungsvollen Quellen
Paper from responsible sources
FSC® C105338

If you have any concerns about our products,
you can contact us on
ProductSafety@springernature.com

In case Publisher is established outside the EU,
the EU authorized representative is:
Springer Nature Customer Service Center GmbH
Europaplatz 3, 69115 Heidelberg, Germany

Printed by Libri Plureos GmbH
in Hamburg, Germany